Student Laboratory Manual

Biology:
A Search for Order in Complexity

Second Edition

Original Title: *Investigations into Biology, A Student Laboratory Manual*
Original copyright © 1973 by The Zondervan Corporation, Grand Rapids, Michigan
Copyright to the first edition transferred to Harold S. Slusher in 1991

Student Laboratory Manual, Copyright © 2005 Christian Liberty Press
for *Biology: A Search for Order in Complexity, Second Edition*
Copyright © 2004 by Christian Liberty Press Copyright

A publication of
Christian Liberty Press
502 West Euclid Avenue
Arlington Heights, Illinois 60004
www.christianlibertypress.com

Written by: Robert C. Glotzhober
             John M. Fricke
             Laurence L. Meissner,
             John N. Moore
Layout and editing: Edward J. Shewan
Copyediting: Diane C. Olson

ISBN 1-930367-94-5

Printed in the United States of America

CONTENTS

To the Student

Unlike history, English, and many other courses that you may take in school, biology should involve much more than just reading a book, discussing it with other students, or listening to a lecture. **Biology** (from *bios*, life; and *logos*, word or study) involves *the study of living things*. Although much can be learned from reading the text and hearing lectures, much more must be learned by looking at the animals and plants themselves. There is no substitute for seeing the "real thing." For this reason, the authors of this lab manual have tried to prepare laboratory exercises that will enable you to see, question, and learn in first-hand experiences.

In preparation for each lab exercise, you should fully and carefully read through the material before beginning each lab. If you do this preliminary reading, the work will be easier for you, and you will be able to work faster and learn more from the exercise. Some questions may be answered only by doing the experiments, whereas answers to other questions may be found in the text. Still other questions require original thinking on your part and give you a chance to offer your opinion.

Most important from the authors' viewpoint, this lab manual emphasizes good biological principles, and the glories of the handiwork of God in His marvelous creation are made evident. The power and wisdom of God can be clearly seen in nature and it is our hope that in your examination of nature, you will look for His handiwork and learn to enjoy, marvel at, and appreciate it.

About the Authors:

Robert C. Glotzhober, a former biology teacher, received his M.S. degree from Michigan State University at East Lansing. He is presently curator of Natural History at the Ohio Historical Society.

John M. Fricke earned his M.A.T. degree from Michigan State University at East Lansing. He is a former biology teacher and presently a professor of biology at Concordia University, Portland, Oregon.

Laurence L. Meissner, received an M.S. degree at Eastern Michigan University, Ypsilanti, is presently a professor of biology at Concordia University, Austin, Texas.

John N. Moore, M.S., Ed.D., was the original project coordinator. In addition, Dr. Moore was one of the original founders of the Creation Research Society and a former managing editor of the *Creation Research Society Quarterly*. He edited the first creationist textbook *Biology: A Search for Order in Complexity*. He team taught a natural science course at Michigan State University with an evolutionist.

EXERCISE 1-1
THE SCIENTIST AND HIS WORK

INTRODUCTION:

Very often the scientist observes the results of some event without actually seeing the event itself. For example, we can find many fossils and yet we did not observe the fossil in the process of formation. Reasoning logically, we can describe the process of fossilization, even though we do not directly see it happening. Therefore, we are drawing a conclusion on the basis of indirect evidence. Let's try an experiment in which we use only indirect evidence and see what conclusion we can suggest.

PROBLEM:

To determine how many sides of a sugar cube are marked.

MATERIALS:

One quart milk carton for each student (or any similar tall, opaque container); one sugar cube with an unknown number of sides marked with a spot.

PROCEDURE:

A. Observation

Observe the sugar cube in the container and determine whether or not a spot is visible on the upper face of the cube. Record your observation on the chart on page 2. Now shake the container and observe the cube again. Record your observation as above. (Do not look into the container while it is being shaken.) Repeat this process until 100 trials are completed and recorded.

B. Conclusion

We have looked at 100 faces of one cube shaken 100 times. A cube has 6 faces or sides. If in the 100 trials we observed 50 spots, then half of the faces we saw had a spot (averaging 0.5 spots per face). It would seem logical that half of the faces of the cube have a spot on them. The total number of sides with spots can be computed by multiplying the spots per side times the number of sides on the cube: (0.5 x 6 = 3).

RECORD OF OBSERVATIONS

Check after each trial whether spot was observed or not observed.

	no	yes		no	yes		no	yes		no	yes		no	yes
1			21			41			61			81		
2			22			42			62			82		
3			23			43			63			83		
4			24			44			64			84		
5			25			45			65			85		
6			26			46			66			86		
7			27			47			67			87		
8			28			48			68			88		
9			29			49			69			89		
10			30			50			70			90		
11			31			51			71			91		
12			32			52			72			92		
13			33			53			73			93		
14			34			54			74			94		
15			35			55			75			95		
16			36			56			76			96		
17			37			57			77			97		
18			38			58			78			98		
19			39			59			79			99		
20			40			60			80			100		

DISCUSSION:

1. How many spots did you observe in your 100 trials? _____

2. What is the average number of spots per face? _____

3. From your observations, how many spots do you conclude are on your cube? _____

4. If you did only 10 trials, how would this affect your results? _____

   _____

5. Would you come up with more or fewer spots on your cube? _____

6. Were there any series of 5 trials in which no spots were observed? _____

7. If you used only 5 trials similar to the above situation, what would be your results?

   _____

   _____

8. How many trials are necessary to get enough data to have reliable results? _____

    _____

9. How many faces of your cube really had a spot, and why was this different from the
    number you obtained from your observation? _____

    _____

    _____

EXERCISE 1-2
LIVING AND NONLIVING THINGS

INTRODUCTION:

Living things are unique. They alone carry on life processes that enable them to survive and reproduce after their own kind. But living things are also composed of the same basic materials that form nonliving things (matter, elements, compounds, and mixtures).

PROBLEM:

To determine the characteristics that living and nonliving things have in common and what characteristics are most typical of living things.

MATERIALS:

Text; petri dishes; sodium chloride crystals; 1% silver nitrate solution ($AgNO3$); medicine dropper.

PROCEDURE:

A. Characteristics of Living Things

List the characteristics you would find in a living thing. Can you come up with 10?

_____   _____

_____   _____

_____   _____

_____   _____

_____   _____

B. Characteristics of Nonliving Things

List some nonliving things and tell what characteristics each has in common with a living thing.

_____

_____

_____

_____

C. Precipitation of Silver Chloride

Pour just enough silver nitrate solution into a petri dish to cover the bottom. Then add a few grains of sodium chloride crystals. Record your observations (*top of next page*).

_____

_____

Are these materials living or nonliving? Explain. _____

_____

_____

D. **Observation**

Your teacher will show you a "thing" that seems to show living characteristics.

What are these characteristics?_____

_____

_____

Is this "thing" really alive? Explain. _____

_____

_____

DISCUSSION:

Explain in a paragraph how you would distinguish between living and nonliving things. _____

_____

_____

_____

_____

_____

_____

EXERCISE 2-1
PROTEIN DIVERSITY

INTRODUCTION:

You have learned that proteins are very special organic compounds. Proteins are important parts of cell structures such as the cell membranes, ribosomes, and golgi bodies; hair, skin, and nails; and important parts of all enzymes. We also know that there are hundreds of thousands of proteins, and yet chemical analysis of all of these proteins shows they are composed of approximately twenty different amino acids. In this activity we will see how with a limited number of amino acids we can come up with many small protein-like molecules.

PROBLEM:

To determine why many proteins can be formed from a few amino acids.

MATERIALS:

Diagrams of various amino acids; molecular model kits.

PROCEDURE:

A. Peptide Bond

Given below are two simple amino acids. In the process of protein synthesis, amino acids are linked together, as shown:

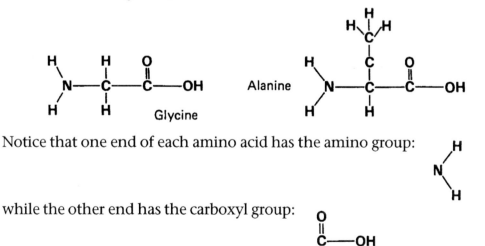

Notice that one end of each amino acid has the amino group:

while the other end has the carboxyl group:

Before a peptide chain or a protein can be formed, individual amino acids must be joined together. This requires formation of a special chemical bond called a peptide bond. Let's see how this bond can be formed.

The peptide bond is formed like this: The hydroxide (—OH) of the carboxyl group of one amino acid and one hydrogen from the amino group of the other amino acid are removed by enzymes. This hydroxide and hydrogen combine to form water. The remains of the carboxyl and amino groups are brought together. See this step visually on page 8.

7

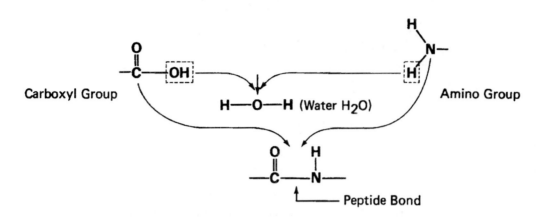

Carboxyl Group   H—O—H (Water H₂O)   Amino Group

Peptide Bond

Now we'll show the peptide bond formed along with the other parts of the two amino acids.

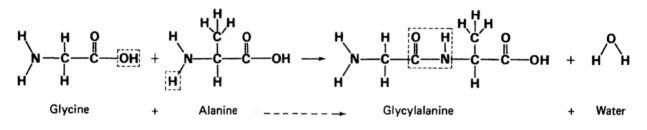

Glycine   +   Alanine   - - - - - - - →   Glycylalanine   +   Water

Here are another two amino acids. Show how these could combine to form a small chain, using the same method used above.

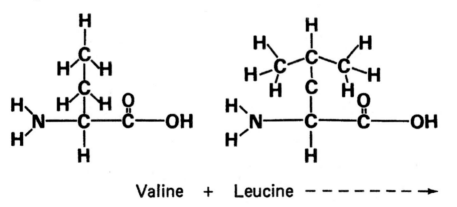

Valine   +   Leucine   - - - - - - - - →

B. Peptide Chains

Now let's go one step farther to try to determine why so few kinds of amino acids are necessary for producing thousands of proteins. Use the four amino acids given so far in this lab exercise, letting the first letter of each one stand for all of the amino acids except the amino group and the carboxyl group.

For example, glycine would be shown as:

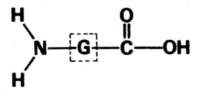

Draw as many different kinds of peptide chains as possible, making your chains only four (4) amino acids long, and never using one amino acid more than once in a chain.

DISCUSSION:

What seems to be an explanation of why there are so many proteins? _____

_____

_____

_____

_____

_____

_____

_____

EXERCISE 2-2
ENZYMES AND THEIR FUNCTIONS

INTRODUCTION:

If you have ever tried to burn sugar in a spoon over a Bunsen burner,[1] you probably have noticed that a great amount of heat was released as the sugar burned. If sugar burned like that in your body, too much heat would be produced, damaging your body's cells and tissues. Yet sugar must be burned or converted into carbon dioxide (CO_2) and water H_2O to provide energy for your life processes.

The chemical changes which are necessary to break down sugar without too much heat being produced are controlled by a special group of compounds called enzymes. So far, scientists have identified at least 1,500 of them, and many more are being studied. In this lab we will study diastase. This enzyme comes from barley malt and can change corn starch into glucose (a sugar).

PROBLEM:

To determine the optimum conditions for the activity of diastase (a mixture of amylases from malt)[2] in the digestion of corn starch.

MATERIALS:

Three percent (0.3%) solution of corn starch; Lugol's iodine staining solution[3]; 0.2% solution of diastase; Benedict's solution[4]; hot water bath test tubes; glass markers.

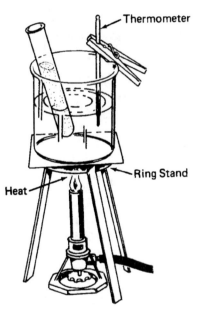

WATER BATH

PROCEDURE:

A. **To Determine at What Temperature the Maximum Rate of Digestion Occurs**

Fill 6 test tubes 3/4 full of 0.3% corn starch solution. Add 1 drop of Lugol's iodine solution to each test tube. Number the test tubes 1 to 6 with a glass marker. Place one test tube in each of the following water baths: 10 C; 20 C; 30 C; 40 C; 50 C; and 60 C. (See illustration at right.)

Now add 5 drops of the 0.2% diastase solution to each test tube, recording on the chart on the next page the time that you add it. When the solution in a tube becomes clear,

---

1. A Bunsen burner is a device used in chemistry for heating. Robert Wilhelm Bunsen did not invent this device but, in 1855, improved its design to aid his work in spectroscopy. Instead of mixing the gas with the air right at the point of combustion (i.e., Michael Faraday's design), he proposed that the gas should be mixed with the air *before* combustion.

2. Any of a group of enzymes (as amylopsin) that catalyze the hydrolysis of starch and glycogen or their intermediate hydrolysis products.

3. Lugol's iodine solution is 5 gm iodine with 10 gm KI per 100 ml of solution, yields 6.3 mg iodine per drop. Since 5 gm/100 ml and 10 gm/100 ml approximate a 5% and 10% concentration, it appears that "strong iodine solution" and Lugol's solution are essentially the same.

4. Benedict's solution is a deep-blue alkaline solution used to test for the presence of the aldehyde functional group (–CHO). The substance to be tested is heated with Benedict's solution; if a brick-red precipitate forms, the presence of the aldehyde group is verified.

record the time again on the chart. Remove each test tube from the water bath, and test for sugar by adding Benedict's solution and boiling (be careful to move the test tube around rapidly, and do not point the end at anyone). If an orange or red color results, sugar is present.

Tube	1	2	3	4	5	6
Temperature						
Starting time						
Time clear						
Elapsed time						
Sugar present						

Construct a graph of your results, putting the time required for digestion of the starch on the vertical axis and the temperature on the horizontal axis.

B. To Determine the Effect of Concentration on the Rate of Digestion

Fill 6 test tubes 3/4 full of corn starch solution and add 1 drop of Lugol's iodine to each one. Number the test tubes 1 to 6 with a glass marker. Add diastase solution to each test tube, doubling the amount added to each tube. Add 1 drop to the first tube, 2 drops to the second, 4 to the third, 8 to the fourth, 16 to the fifth, and 32 to the sixth. Place all of the tubes in a warm water bath of approximately 30° C; then record on the chart below the time required for digestion to be completed (completed when clear).

Tube	1	2	3	4	5	6
Starting time						
Completion time						
Elapsed time						

Make a graph on a separate sheet of paper showing the relationship between elapsed time and the concentration of diastase solution, placing time on the vertical axis and concentration on the horizontal axis.

DISCUSSION:

1. What does Lugol's solution test for, and how is this shown?_____

   _____

2. What does Benedict's solution test for? _____

   _____

3. What is the general effect of temperature on the rate of digestion?_____

   _____

4. Is there an optimum temperature for digestion by diastase?_____ What is it if there

   is one? _____

5. What is the general effect of concentration of the enzyme on digestion? _____

   _____

EXERCISE 3-1
THE MICROSCOPE AND ITS USE

INTRODUCTION:

The science of biology had its greatest impetus when the microscope was invented. Through the microscope, we see into a world unknown to man before the sixteenth century. Before that time, man could recognize only those organisms that were visible to the unaided eye, and he could only imagine living things smaller than the eye could see. When using the microscope, remember that it is an expensive precision instrument. With care it can serve you for a lifetime.

PROBLEM:

To become familiar with the parts and operation of a microscope.

MATERIALS:

Compound microscope; slides; cover slips; newsprint; water; medicine dropper; lens paper.

PROCEDURE:

A. **Getting Acquainted with Your Microscope**

Handle the microscope that you are going to use with great care—one hand under the base and the other hand firmly gripping the arm. Set your scope well away from the edge of the table. With lens paper, clean all dust and fingerprints from the lenses and the mirror. NEVER CLEAN THE LENSES WITH ANYTHING BUT LENS PAPER! The lenses are actually quite soft and can be scratched even by tissue or cloth.

Locate and identify by name all parts of the microscope, referring to the drawing on the next page.

Locate the *nosepiece*. Notice that it has 2 (or more) objectives mounted on it. Rotate the nosepiece so that the first objective lens is lined up with the *ocular lens*; then the next; and so forth. Notice how the objective lenses click into place. Put the low-power lens (10X or 12X) into position, lined up with the ocular lens. Now turn the *coarse adjustment* and observe which way the tube moves. Also turn the *fine-adjustment* knob. CAUTION: Never turn the fine adjustment knob while holding the coarse-adjustment knob. How do the movements of the lenses compare when using the coarse and fine adjustments? _____

_____

With the scope on low power and the objective lens in place, "rack" the tube all the way down while looking from the side. Look through the eyepiece and adjust the mirror to get a gray-white field of light. While looking through the eyepiece, adjust the diaphragm to control the amount of light.

Note: Whenever you "rack down," look from the side.

B. Making and Observing a Slide

From a piece of newsprint, cut out a letter "e" and place it in the center of a slide. Add a drop of water and cover with a cover slip. Place this slide on the stage of your scope and clamp it in place with the stage clips. The letter "e" should be positioned in the middle of the objective lens. With your scope on low power, look from the side and rack your scope down, being careful not to rack it too far, onto or through the slide itself. Notice how close the objective comes to the slide.

This compound microscope is a typical optical microscope found in many high school laboratories.

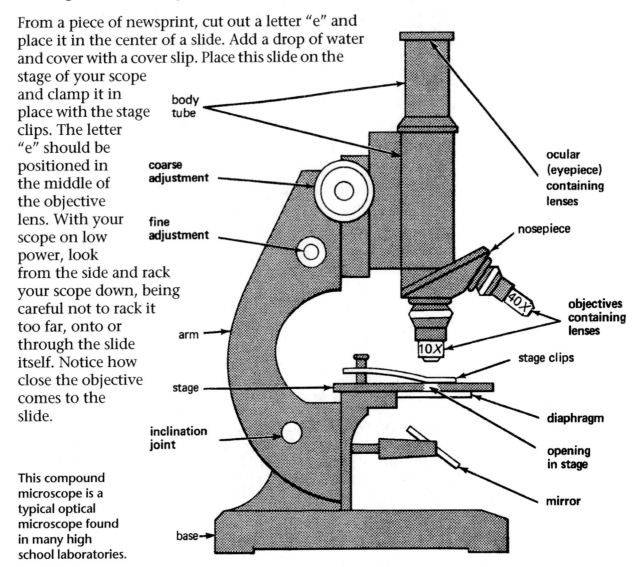

BODY TUBE:	Holds lenses of eyepiece and objectives at correct distance from each other.	**OCULAR:**	Has lenses to increase magnification. They may be replaced by others of different magnifications.
COARSE ADJUSTMENT:	Moves body tube up and down in relation to specimen.	**NOSEPIECE:**	Permits interchange of high- and low-power objectives.
FINE ADJUSTMENT:	Permits precise focusing by moving body tube or stage up or down slightly.	**OBJECTIVES:**	Have lenses of different magnifications. In this case, the low-power objective is 10X; the high-power, 40X.
ARM:	Supports the body tube and the coarse adjustment.	**STAGE CLIPS:**	Hold slide in place.
STAGE:	Supports the slide.	**DIAPHRAGM:**	Regulates light passing through specimen.
INCLINATION JOINT:	Permits tilting of microscope to adjust to eye level.	**OPENING IN STAGE:**	Admits light.
BASE:	Supports microscope.	**MIRROR:**	Reflects light upward through opening in stage and specimen.

Looking through the eyepiece, adjust the mirror and diaphragm to get the best light. Slowly rack your scope upward until the newsprint letter is in focus. If you racked it too far up, turn it all the way down (while watching from the side) and then start over again, moving it more slowly as you watch through the eyepiece.

Now, with a thumb on either side of the slide, move the slide to the right. In what direction does the image move? _____

Move the slide away from you. In what direction does the image move? _____

Move the slide so that the image is in the exact center of the field of vision. Draw the letter exactly as you see it in the space provided. Title your drawing and give the power of magnification of the ocular (eyepiece) by that of the objective. (If the ocular is 1OX and the objective is 12X, then the magnification is 120X.)

What is the comparison between the image of the "e" that you saw under the scope and its normal image? _____

___

C. Observing the Slide on High Power

Now we are ready to look at this slide on high power (40X). Rack your scope all the way up on low power, and switch to the high-power objective. Be sure that it has clicked into place. Then rack your scope down as close to the slide as possible without actually touching it. REMEMBER: LOOK FROM THE SIDE! Notice that you can get much closer to the slide with the high power objective, and you must be careful not to break the slide or scratch the lens when racking down.

Focus upward until the object comes into view. If you go too far, rack the scope downward carefully again, looking from the side. When you have focused the image with the coarse-adjustment knob, try to bring the image into finer focus by using the fine adjustment.

Draw in the space provided the image you see on high power. Title your drawing, and give the power of your scope on high power.

D. Making Another Slide

Place threads of three different colors on a slide in such a way that they lie across each other. Observe them on low power and then on high power. Using your fine adjustment, determine what is the order of the threads by color from top to bottom.

E. Clean-up

When you have finished using your microscope for the day, remove the slide from the stage and return it to its proper place. Put the low-power objective in place and rack your scope all the way down. Again clean all lenses carefully with lens paper and return your scope to its proper storage place, carrying it as instructed above.

DISCUSSION:

1. Briefly review your observations and list below what a microscope does to an image.

   _____

   _____

   _____

   _____

2. Why can't all the threads be seen clearly at the same time?

   _____

   _____

   _____

EXERCISE 3-2
CELLS: A COMMON DENOMINATOR

INTRODUCTION:

You may know that most living things are made of cells. Yet to see cells often requires special equipment and procedures. In this lab activity we will take a look at many different kinds of cells and see how they are alike and how they are different.

PROBLEM:

To determine how cells are alike and how they are different.

MATERIALS:

Onion; elodea; slides; cover slips; Lugol's iodine stain; 5% salt solution; medicine droppers; forceps; single-edge razor blades; toothpicks; methylene blue stain.

PROCEDURE:

A. **The Onion Skin Epidermis**

From a quartered onion, take one of its many storage leaves. Notice that one side is concave and the other convex. The cells you are looking for are located on the concave side of this leaf. With a forceps, remove the thin sheet of cells from the concave side of the leaf. Place this sheet of cells on a drop of water on a slide. Be careful to keep the cell layers from rolling up. Cover with a cover slip. Observe under low power and then under high power. You may have to move your slide around to get a clear picture. What cell structures can you see? _____

_____

_____

Remove your slide from the microscope and place a drop of iodine solution at the edge of the cover slip. Draw the iodine to the onion skin material by touching some paper toweling to the fluid at the opposite end of the cover slip. Observe again on low and high power. What is now different about the cells? _____

Where is the iodine stain concentrated? _____

Draw five cells as seen on high power. Label: cell wall, cytoplasm, nucleus, nucleoli.

B. Elodea Cells

From an elodea plant, remove a single leaf and mount on a slide in a drop of water. Cover with a cover slip. Locate a thin section near the edge of the leaf where the cells are only 1 or 2 cell layers thick. Switch to high power and observe. The oval, green bodies are chloroplasts. Are they moving? _____ If they are, describe their movement. _____

_____

Notice that the center of the cell looks clear. It is not really empty, but is filled with a watery cell fluid. Draw and label cells as you see them.

So far, we have not seen evidence of the cell membrane because it is pressed tightly against the cell wall. Using a strong salt solution, we can draw water out of the cell. This will cause the volume of the cell contents to shrink, and the cell membrane will come away from the cell wall. Although the membrane is less than 100 Å (i.e., 100 angstroms, or 100 x 10^{-10} m; or 10 nanometers = 10^{-8} m) in thickness and cannot be seen by the light microscope, you will be able to see the abrupt boundary within which the cytoplasm is confined. To effect this, place a drop of 5% salt solution at one edge of the cover slip, then draw this solution under the cover slip by touching some paper toweling to the water at the other end of the cover slip. Describe what is happening.

_____

_____

Define the term "plasmolysis." _____

_____

_____

Draw and label elodea cells as you saw them after adding the salt solution. Label: cell wall, chloroplasts, nucleus, vacuole, cell membrane.

C. Human Cells

With a toothpick, gently scrape the inside of your cheek and place this material on a slide. Add a drop of water and a drop of methylene blue stain. Allow to stain for one minute. Look for cells first on low power. Center them and switch to high power. Explain how this cell is like the onion skin cell.

_____

_____

How is it different from the elodea? _____

_____

Draw a group of cheek lining cells showing how they fit together. Label: boundary of cell membrane, nucleus, cytoplasm.

DISCUSSION:

1. What characteristics do all of these cells have in common?_____

   _____

   _____

2. What unique characteristics do any of these cells have? _____

   _____

   _____

   _____

3. Not all living organisms are cellular. Do some research study on viruses and slime
   molds and discuss how their structure differs from cellular organizations. _____

   _____

   _____

   _____

   _____

EXERCISE 3-3
MITOSIS

INTRODUCTION:

We all began life as a single cell. Now we are literally trillions of cells. How did one cell become so many? The process by which this happens is called mitosis. In this lab activity we will observe some prepared slides of cells that are undergoing mitosis. Then we will prepare some slides of our own that show mitosis.

PROBLEM:

To determine the steps in mitosis and how we can prepare materials to see the chromosomes and changes occurring in them.

MATERIALS:

Onion sets with young roots; slides; aceto-carmine stain; cover slips; hot water bath; prepared slides of plant and animal mitosis; paper toweling; watch glass; razor blades.

PROCEDURES:

A. **The Stages of Mitosis**

Obtain from your teacher a prepared slide of an onion root tip showing mitosis. Using low power, scan the material on the slide and locate an area near the tip of the root where the cells are all very small. This will be an area where mitosis was occurring when this slide was made. Now switch to high power. Cells that are undergoing mitosis will have nuclear structures that are not visible in other cells that are not dividing. The first stage of cell mitosis is *prophase*. Look for cells in which the nuclear membrane has disappeared and the nucleoplasm of the nucleus appears like a tangled mass of threads. These threads are called chromosomes. In the next stage, called *metaphase*, the threads become shorter and thicker and are lined up at the center of the cell, called the *equator*. In *anaphase*, the paired chromosomes (more accurately chromatids) separate, with one complete set going to each pole. At *telophase*, there is a set of chromosomes at each pole. A new cell wall begins to form, and the chromosomes become enclosed by a nuclear membrane, thus forming two daughter cells. These two new cells may enter into a nonmitotic phase, called *interphase*.

Onion Root Tip

21

Make drawings of the stages of mitosis from this prepared slide. Label: chromosomes, nuclear membrane, centrosome, spindle, daughter cells.

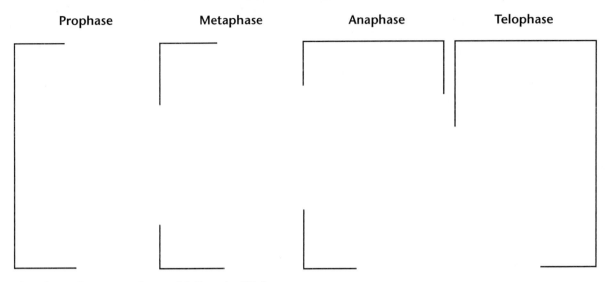

| Prophase | Metaphase | Anaphase | Telophase |

B. Student Preparation of Mitosis Slides

Remove the lower 1/2 inch of a young root from an onion plant. Place this in a watch glass filled with aceto-carmine stain. Heat the stain and root tips over a hot water bath for 5 minutes. Remove the tip from the stain and place it on a clean slide with a drop of fresh aceto-carmine stain. With a razor blade, remove and discard all but the most deeply stained part of the root tip. Now cut the remaining portion into pieces about the size of a pin head. Place a cover slip over the material on the slide. Using your thumb, press downward on the cover slip firmly to mash the cells into a thin layer.

Note: Be careful not to allow the cover slip to move sidewise, and remember that the cover slips break easily.

Remove excess fluid from the slide and observe on low power. Look for cells with deeply stained threads (the chromosomes). Now switch to high power and locate the various stages of mitosis. Present your slide to the teacher for observation and comment.

DISCUSSION:

What is the importance of the chromosomes' duplicating and dividing during mitosis?

_____

_____

_____

_____

_____

EXERCISE 3-4
HUMAN INHERITANCE

INTRODUCTION:

People have been concerned about human genetics for a long time. Once, it was very important to know your family ancestry for social reasons. Today we are concerned about human genetics to try to predict and prevent hereditary diseases such as hemophilia (bleeder's disease) and the problems that result from the Rh factor in blood (see section 7-18 of the textbook). Many of the human genetic characteristics are complex, and some we do not know much about. There are several human traits, however, that we can easily observe and study.

PROBLEM:

To discover the functioning of genetics in several simple human inherited characteristics.

MATERIALS:

PTC test paper, used to test for the genetic ability to experience a bitter taste; a dominant taster gene allows 70% of the population to taste PTC.

PROCEDURE:

A. PTC Test Paper

Obtain a small piece of PTC paper from your teacher. Place this in your mouth and chew on it for several seconds. PTC paper is treated with the compound Phenylthiocarbamide. To some people, this substance has no taste; to others, it tastes bitter (or in some cases salty, sweet, or sour). The ability to taste PTC is an inherited trait. Test several people and record your results in the chart on the following page, indicating with a "+" for each one who tasted it, and a "–" for those who did not. Total the number of tasters and nontasters you tested and record this as well.

B. Earlobe Shape and Tongue Curling

Some other genetic traits that are easy to study are earlobe shape and tongue curling. Some people have earlobes that are attached, while others are unattached (Figure 1). Likewise, some people can curl their tongue into a tube (Figure 2) and others cannot. Test yourself and several others for these traits and record the results on the chart.

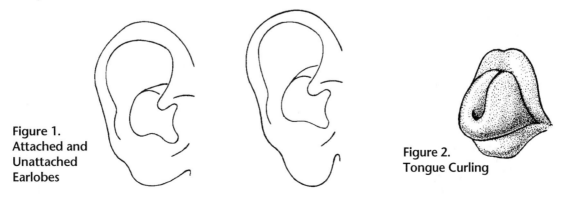

Figure 1.
Attached and
Unattached
Earlobes

Figure 2.
Tongue Curling

Chart of Your Observations					
Characteristic	Yourself	Father	Mother	# of Tested +	# of Tested –
PTC Paper Tasting					
Tongue Curling					
Earlobe Shape[a]					

a. For attached earlobes use a "+" sign, and for unattached earlobes use a "–" sign.

DISCUSSION:

1. What percentage of those tested taste PTC? _____

   What percentage of them can curl their tongue? _____

   What percentage have unattached earlobes? _____

2. In general, 70% of the population can taste PTC, 66% can curl their tongue, and 75% have unattached earlobes.

   Does the average of those tested match these ratios perfectly? _____ If not, explain why not. _____

   _____

3. Tasting of PTC is a dominant trait. If both parents of a person are tasters, is it possible that he could be a nontaster? _____ Explain. _____

   _____

4. Unattached earlobes is a dominant trait. A family has four children, two of whom have attached earlobes and two, unattached. What would be the most probable genotype of the two parents? _____

   _____

5. If both parents are heterozygous for tongue curling, can all of their four children be noncurlers? _____

   Explain. _____

   _____

6. On the chart above you recorded the phenotype of both your own and your parents' traits. Can you compute the possible genotype for these? _____

   _____

   _____

EXERCISE 3-5
GENETIC CROSSES WITH FRUIT FLIES

INTRODUCTION:

Why are some people tall and others short? How is it possible that two tall parents can have a short child? These are only two of the many questions that you may ask about genetics. Much of what is known today about genetics has been learned as a result of work done with the common fruit fly, Drosophila melanogaster. It is small and reproduces readily, producing a large number of offspring in a short time. It also has many traits that are easily studied, and yet has only 4 pairs of chromosomes.

PROBLEM:

To determine the manner in which traits are inherited in the fruit fly.

MATERIALS:

Stock cultures of a wild type of fruit fly; stock cultures of vestigil-winged fruit flies; 1/2-pint culture jars for crosses; ethyl ether; etherizer; cotton plugs for jars; pressure cooker; camel's hair brush; petri dish; filter paper; nutrient agar; cornmeal; corn syrup or molasses; active dried yeast; mold inhibitor; morgue (jar of light motor oil with lid).

PROCEDURE:

A. Culture Media

To raise fruit flies, we must have a culture on which they can mate and lay eggs, and in which the larvae can tunnel. Weigh out 15 grams of agar and add it to 500 ml of water. Heat this to a boil, then slowly add 135 g of corn syrup or molasses. Mix 100 g of cornmeal in 250 ml of cold water and add this to the above mixture. Boil slowly for 5 minutes and then add a drop or two of mold inhibitor. Pour this into your sterilized culture bottles until they have a layer 1/2 to one inch deep on the bottom. (Sterilize culture bottles in advance in a pressure cooker for 15 minutes at 15 pounds pressure. Be careful not to open pressure cooker until all pressure is released.) Cork the culture bottles immediately with cotton plugs. Set the culture on an angle until the agar hardens, as this provides a larger surface area for the flies. These culture jars may be kept refrigerated until needed, but be sure to warm them to room temperature before adding the flies. Also, before adding the flies, add a pinch of dried yeast to the moist surface of the culture. This will provide food for the larvae.

B. Etherization and Transfer of the Flies

Take your stock culture of flies and tap it on the table. This will momentarily knock the flies to the bottom. Quickly remove the top of the container and cover it with the etherizer. Now invert the containers and tap again so that the flies fall into the etherizer. Quickly recap the bottle. As soon as you see that all of the flies are etherized, dump them out of the etherizer onto a white card for sorting. If you leave them in the etherizer too long, or if they come into contact with the ether, it will kill them. Dead flies are recognized by their wings standing out at an angle.

The flies should remain etherized for about 5 minutes, but if some of them begin to recover too soon, you will need to re-etherize them. To do this, tape a piece of cotton or filter paper to the bottom of a petri dish, add ether to it and invert it over the flies for a moment.

When adding etherized flies to a culture bottle, set the bottle on its side and brush the flies off the white card into the bottle. If the flies fall into the culture, they may stick to its surface and die.

Sometimes students find that tapping a jar does not work to get the flies into the etherizer. You may choose, instead, to place a light bulb above the culture bottle and etherizer, letting the flies be attracted upward towards the light and into the etherizer.

FIGURE 1.
Etherization Processes

C. Identification of the Flies

For your experiment you will need to be able to recognize and separate male from female flies, and wild type of winged flies (normal) from vestigial-winged flies.

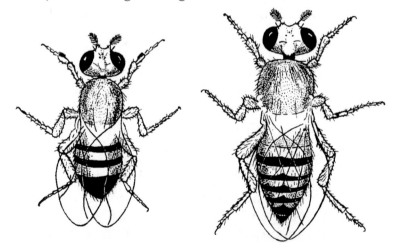

FIGURE 2. Male-female Identification

Females tend to be larger than the males. They have longer and more pointed abdomens with several small rows of black lines across the top of it. Males have more rounded abdomens with a thick black zone near the posterior end of the dorsal surface. Males may also be identified by a black sex comb near the first joint of their front legs.

Wild-type flies (both male and female) have long wings, whereas vestigial flies have only small, stubby, and irregularly shaped wings.

Etherize a culture of wild-type flies and learn their characteristics, and the distinctions between male and female. Separate the males from the females on your white card by pushing them to opposite sides with the camel's hair brush. When you have finished, check your accuracy with your teacher. Repeat the same with the culture of the vestigial-winged flies.

D. The Cross

Female fruit flies are capable of storing sperm in their bodies for a considerable time after mating. Before you make your cross, you must be sure that you are using virgin females, or else they will lay eggs that were fertilized with sperm from a male in their stock culture, and not by the male in the cross. An experienced worker can usually separate the virgin females from others by sight, but this skill is not quickly learned.

The easiest way to obtain virgin females is to remove all adult fruit flies from your stock culture about 10 hours before making your cross. Females usually do not mate until 12 hours after hatching. Therefore, after a 10-hour wait, all females in the culture should be virgin.

Students' groups should now be divided in half. Half of the groups should cross males from the wild type of stock with females from the vestigial stock, and the other half of the groups should do the reverse. Each new culture bottle should be started with about 5 males and 5 females (no fewer than 3 of each). Label the jar to include the date and the types and sexes of flies crossed.

The female fruit flies should begin laying eggs within 36 hours. These eggs hatch quickly and soon larvae begin tunneling through the media, eating the yeast cells. After several days, the larvae enter the pupal stage. About 12 days after mating, these pupae begin to hatch into adult flies. Before this time (at about the tenth day), you should remove the parent flies. This avoids unwanted recrossing.

Once the flies begin to emerge, you should etherize and remove the new flies (F_1 generation) each day for about a week to 10 days. Separate them with your camel's hair brush as to (1) wild type *and* (2) vestigial-winged. Record your daily counting, and dispose of any flies that you don't need for making new crosses by putting them into the "morgue" (the jar with oil in it).

E. Producing the Second Generation

When you first start obtaining young flies of the F_1 generation, repeat the methods for crossing flies, and start a new culture, using males and females both from the F_1 generation produced in step D. Record the types of offspring from this cross (the F_2 generation) the same way as you did for the F_1.

DISCUSSION:

1. How many of each type of flies (wild-type *and* vestigial-winged) were produced in the F_1 generation?_____

   _____

2. If possible, check with the results of other students. Were there any differences in the F_1 generation between crosses of wild-type males *and* vestigial females and those crosses with vestigial males *and* wild-type females? _____

   _____

   _____

3. How many of each type of flies were produced in the F$_2$ generation? _____

   _____

4. Compute the answer to number 3 in the form of a percentage. _____

5. Which trait appears to be dominant—wild-type wings or vestigial wings? _____

   _____

6. Diagram each cross, showing first the genotype of each parent, and then the possible ratio of genotypes of the offspring.

7. Did your cross ratios match the ratios expected from your figures in number 6 above? _____ Explain. _____

   _____

   _____

8. Explain what would happen to your ratios if you used only the first 20 flies from your data to compute the ratio of offspring. _____

   _____

   _____

EXERCISE 3-6
FERTILIZATION OF SEA URCHIN EGGS

INTRODUCTION:

The workings of genetic inheritance depend upon a mixing of the available gene types. This mixing and recombination of these genes takes place in meiosis and fertilization. Fertilization, therefore, is the instantaneous occurrence that determines what traits an organism will inherit. In many plants and animals, fertilization is also necessary to start the life of the organism. It is a wonderful and amazing process, as well as biologically important.

PROBLEM:

To observe the actual fertilization of sea urchin eggs and to study their early development.

MATERIALS:

Live sea urchins; filtered sea water; clean or sterile glassware; medicine droppers; petri dishes; hypodermic syringe; 100-ml graduated cylinder beakers; glass-marking pencil; 0.5 molar solution of potassium chloride (KCl); depression slides; cover slips; microscope.

PROCEDURE:

A. Obtaining Eggs and Sperm

Prepare several clean and sterile beakers and petri dishes. The beakers should be of a size that will enable you to support the sea urchin over the mouth of the beaker (usually 250-ml beakers are satisfactory). Fill each of the beakers with 25 ml of sea water. Keep the petri dishes dry.

Now inject about 2 ml (2 cc) of the potassium chloride solution into several sea urchins through the soft membrane surrounding the ventral mouth. Place the urchins upright on a clean surface until they begin to shed their gametes. This will take only a few minutes. A male's sperm are milky white, while a female's eggs are a creamy yellow.

Once ovulation (release of eggs) has begun, invert the *females* over the beakers of sea water and let the eggs drip into it. This will produce a concentrated egg solution, as each female can release up to one billion[5] eggs. (The males produce a trillion[6] sperm.) The *males* should be inverted in a *dry* petri dish. In this dry form, the sperm can be covered and kept in a refrigerator up to 24 hours. Once they are added to sea water, however, they will live only about 20 minutes. Unfertilized eggs can be kept about 6 hours in a refrigerator.

Just before you are ready to use the eggs and sperm, prepare dilute cultures of them. For the eggs, add 5 drops of the concentrated culture to 100 ml of sea water. For the sperm, add one or two drops of dry sperm to 10 ml of sea water. Be sure that the sperm are used right away after mixing this culture.

---

5. Note that this 10^9; in the British system, this is called a "milliard."
6. Note that this 10^{12}; in the British system, this is called a "billion."

B. **Observing the Eggs**

Place a drop of the dilute egg solution in a depression slide and put a cover slip over it. Observe this under your microscope, first at low power and then at high power. This will require all of your skills with the microscope, as the eggs are almost transparent. Adjust the diaphragm of the microscope to reduce the amount of light coming through, and the eggs will be easier to see. Near the top of a separate sheet of drawing paper, draw an unfertilized egg as seen under high power.

C. **Fertilization**

Obtain a fresh drop of sea urchin eggs on your depression slide (heat from the light source often kills them, so discard your first drop). To this add one drop of sperm solution and cover this with your cover slip. Immediately place this under the microscope and observe constantly. Within 2 to 5 minutes, fertilization will take place. You should see the small, active sperm swimming all around the eggs. If you watch an egg closely, you should be able to see fertilization actually taking place! Immediately upon fertilization, the egg forms a halo-like *fertilization membrane* that prevents any additional sperm from entering the egg. After you have seen fertilization taking place, draw a picture of a fertilized egg on the drawing paper (under the first drawing).

D. **Development of the Fertilized Eggs**

The chart below gives a timetable for the development of sea urchin eggs. As you can see, the first cleavage takes place in 60 to 70 minutes. If time is limited, your teacher may have fertilized some eggs in advance. Observe eggs that have divided; draw one, labeling it with its approximate age and the temperature at which it was developing. If possible, also observe eggs 4–6 hours old, and draw them.

| Time Schedule for Development of Sea Urchin Eggs ||
Stage	Time After Fertilization
First Cleavage	60–70 minutes
4–8 cells stage	4–5 hours
16–32 cells	6 hours
Early Blastula stage	9 hours
Blastula	11 hours
Late Gastrula	32 hours
Larva form	48–68 hours

To obtain further development, keep egg cultures under 70° F. Each day observe them, and remove any dead or nondeveloping embryos. If necessary, add to or change the sea water after several days. *Never* subject the sea urchins to fresh water. Draw a sketch of your developing sea urchins each day.

DISCUSSION:

1. Why is it important not to put the sea urchins or their eggs in fresh water? _____

_____

_____

2. Why is it necessary for sea urchins to produce such vast numbers of eggs and sperm?

    _____

    _____

3. Why is it necessary to remove dead embryos from the culture? _____

    _____

4. How are dead embryos removed when the eggs are developing in their natural habitat?

    _____

    _____

5. Describe the shape and size of the sea urchin sperm. _____

    _____

    _____

6. Why must the egg divide into two-cell stage, four-cell stage, six-cell stage, and so forth? Why could it not merely grow larger as a single cell, and exist as a single cell in the adult size? _____

    _____

    _____

    _____

    _____

EXERCISE 3-7
FERTILIZATION AND DEVELOPMENT OF FROG EGGS

INTRODUCTION:

In the previous exercise you observed fertilization taking place; you saw some of the first steps of development. It is truly amazing that an entire plant or animal can develop from a single egg. In this lab exercise, we will watch frogs develop from the egg stage to the adult stage, observing the many changes that take place along the way. Frogs are readily available throughout the country, and their changes are easy to observe; hence, they make an exciting lab study.

PROBLEM:

To fertilize frog eggs and observe their developmental stages.

MATERIALS:

Pituitary extract; female and male adult frogs; hypodermic syringe and size 18 bore needle; pond water; petri dishes; frog ringer solution; medicine dropper; microscope and dissecting scope; slides; cover slips; aquarium (*pl.* aquaria).

PROCEDURE:

A. Preparing the Frog

Pituitary extract from the pituitary gland contains hormones that will cause the female frog to ovulate (release her eggs). During a normal annual cycle this takes place automatically in the spring of the year. We can, however, add these hormones artificially at any season of the year and get the same results. This induced ovulation will take from 36 to 72 hours to take effect; so the following procedures must be done several days before the rest of the lab.

Take the pituitary extract and draw it into a syringe with a large bore needle (about size 18). The amount of extract needed will vary from season to season. (In the spring, the extract from only 1 to 3 glands is sufficient, but in the fall it may take as many as 8 to 10 glands to provide enough extract.) Consult your teacher for the amount of extract required. Hold the female frog on her back, pick up the skin of her abdomen, and pull it away from the muscle underneath. Inject the pituitary extract into the cavity between the skin and the muscle layer, being sure not to penetrate the muscles, or to rip open any blood vessels that are just below the skin. As you withdraw the needle, hold your finger over the puncture for a few seconds so that none of the pituitary extract is lost. Place the injected frog into a container with 1/2 to 1 inch of water in the bottom. Keep the temperature of this water below 20° C (preferably at 15° C), as higher temperatures will cause formation of abnormal eggs. After about 2 days, you will notice a few strands of eggs in the water.

B. Fertilization of the Eggs

Once you are sure that the females are ready to ovulate, prepare a solution of sperm as follows. Obtain from your teacher a male frog that has been doubly pithed (both spinal cord and brain removed). From this male, dissect out the white oval testes (ask your teacher for help, and see lab exercise 6–8 on frog dissection). After removing these, rinse them in frog ringer solution and blot them gently on paper toweling to remove blood and body fluids. Cover the bottom of a petri dish with ringer solution and mash the testes in this solution with the end of a glass rod.[7] Let this solution stand for about ten minutes so that the sperm can become activated. Now strip the female of her eggs, by holding her back downward and gently squeezing her abdomen. After stripping several eggs, throw them away and wipe the cloaca dry. Now strip about 100 eggs into each petri dish with the sperm solution. Bathe the eggs continuously for 10 minutes with the sperm solution using a pipette (pī•pĕt′), or medicine dropper. Examine some of the frog eggs and sperm solution during this first 10 minutes to see of you can locate the sperm, and perhaps to see the fertilization membrane forming.

Drain off all the sperm solution and bits of testes after 10 minutes, and flood the eggs with pond water. For the next 15 minutes, a jelly-like substance surrounding the eggs will expand. After this is formed, it is safe to handle the eggs. You may cut them into strips of 10 or so and put these strips into separate petri dishes with clean pond water.

Notice that half of each egg is black and half is white. The black area is the embryo part of the egg; the white is the yolk which will nourish the embryo as it develops. Which of these two areas is oriented up and which is down? _____

_____

Keep most of your containers at room temperature, but take at least one of your petri dishes with the eggs in it and place it in the refrigerator or other cool place.

Examine the eggs under the microscope or dissecting scope several times during the next day and as often as possible for the next week or so. Each time you examine them, be sure to throw away any dead eggs, recognizable by their discoloration or decomposition of tissues. Examine the refrigerated eggs also from day to day as you examine the other eggs.

When the eggs begin to hatch, it is wise to place a sprig of elodea (ĭ•lō′dē•ə, small submerged aquatic herbs) in with them to provide both oxygen and food for the tadpoles. Make sure that the elodea has light during the day. When the tadpole stage is fully reached, place the tadpoles in an aquarium and watch them metamorphose into adult frogs over a period of several weeks or more. Be careful at all stages to remove dead tadpoles, as they will contaminate the entire culture if they are not removed. Be sure also to keep the water clean and oxygenated. Generally only a very small percentage of the original eggs will make it to the adult stage. Your care and attention during their development will determine to a large extent just how successful you will be.

Once the frogs reach adult stage, they will have to have some way of crawling out of the water. Place a floating object in the water (be sure it floats very low or even partially submerged so they can get onto it) or slant a rock, a board, or a plate of glass in such a way as to provide a ramp for them to climb onto.

---

7. Four testes (from 2 males) macerated in this manner, when added to 50 ml of ringer solution, will be sufficient to fertilize about 1,000 eggs (provided by one female frog). This will serve about 8 to 10 students.

DISCUSSION:

1. What changes take place during the first half hour after fertilization? _____

   _____

2. Why is it necessary to separate the eggs into small groups after fertilization? _____

   _____

3. What effect does a lower temperature have on the development of the frog embryos?

   _____

4. What type of environment do tadpoles need to live in? _____

   _____

5. What structure does a tadpole use to breathe with? _____

   _____

6. Where is the mouth in a tadpole? _____

7. What type of feeding is this mouth adapted for? _____

   _____

8. Note the changes that take place in the tail of the tadpole. What functions might this
   organ serve? _____

   _____

9. Why is it necessary to provide a place for the tadpoles to crawl out onto as they
   develop? _____

   _____

10. What changes have taken place in the location, structure, and feeding adaptations of
    the mouth as the tadpole developed into an adult frog? _____

    _____

    _____

    _____

EXERCISE 3-8
DEVELOPMENT OF THE CHICKEN EGG

INTRODUCTION:

Fertilization of an egg is indeed a marvelous occurrence to observe, as was done in exercise 3–6. The big question still remains, however: how does that single-celled egg develop and grow into an entire animal? How can one cell become thousands of cells—bone cells, muscle cells, nerve cells, and many others? What causes and controls these changes is still a mystery to scientists, but they do know what changes take place and when. The study of embryology still promises many fascinating discoveries.

PROBLEM:

To observe and study the development of a chicken egg.

MATERIALS:

Fertilized chicken eggs (2, 6, 12, and 18 days incubated time); thermometer; incubator; saline solution; finger bowls; dissection tools; hand lens; stereoscope; microscope and slides.

PROCEDURE:

A. Incubation

Fertilized chicken eggs can be obtained from a local hatchery. Since we want to observe eggs at different stages of incubation, we will need to stagger their introduction into an incubator. Keep at 50–55°F those eggs that are not incubating (store no longer than a week to 10 days). At the appropriate number of days *before* this lab exercise, place them into an incubator. The temperature should be kept at 100°F. Check this regularly with a thermometer kept at the same level as the eggs. The best development occurs if the relative humidity is near 60%; so keep a shallow dish of water standing in the incubator to maintain that amount of humidity.

The fertilized eggs will need to be turned twice a day. If you mark an "X" with a pencil on one side of the egg, it will be easier to keep track of which eggs have been turned.

B. Examining the Eggs

1. Before you begin dissecting a chicken egg, examine each tool in your dissection kit.

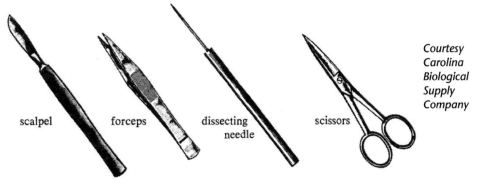

scalpel   forceps   dissecting needle   scissors

Courtesy Carolina Biological Supply Company

2. **Two-day and six-day eggs:** Remove an egg incubated for each of these time intervals from the incubator and carry it to your workspace, being careful to hold it in the same position that it was in when you removed it from the incubator (if you turn it, it will be harder to find the embryo). Set the egg in a finger bowl. Trace a line with your pencil around the top side of the egg, scribing an oval, as shown in the accompanying figure.

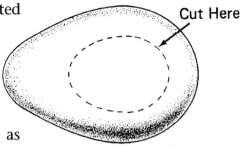

Cut Here

Insert the point of your scissors carefully through a spot along this line, then gradually cut completely around this line. With your forceps, remove the loose piece of the shell.

You should now see the *embryo* floating on top of the yolk. If you do not, locate the fuzzy, twisted cords that connect each end of the egg shell to the yolk, and pull gently on one until the yolk rotates and the embryo is on top. With a medicine dropper, draw off the sticky albumen—the white of the egg. Albumen is not part of the fertilized cell of the egg, but is an accessory substance that appears after fertilization (as does the shell). It is mostly water and protein. What is its function? _____

_____

The yolk is part of the original single cell of the fertilized egg. What is the yolk used for?

_____

_____

Examine the embryo under a hand lens or stereoscope, or if the embryo is still small enough, place it on a slide for study with a microscope. Try to locate the *brain*, the large bulging *eyes*, the *heart* (you may see it beating), and the *spinal cord*. On each side of the spinal cord you may be able to see blocks called *somites* that will form into the vertebrae and muscles.

By the sixth day, you should be able to locate several important membranes. The membrane surrounding the yolk is the *yolk sac*. What structures do you find in the yolk sac?

_____

_____

_____

The *allantois* is a sac emerging from the posterior of the embryo, and is responsible for containing the embryo's waste products. The *amnion* is a membrane surrounding the entire embryo. Push gently on this with a blunt probe. Does it contain anything other than the embryo? _____ What? _____

_____

_____

What is the function of the membrane? _____

_____

_____

To observe the chick embryo more closely, you may wish to carefully cut away the amnion. Compare the development of each egg and record all observations below.

Structure	2 Days	6 Days	12 Days	18 Days	Comments
Development of the Chicken					
Eyes					
Heart					
Brain					
Wings					
Legs					
Feathers					
Beak					
Egg Tooth					
Albumen					
Yoke Sac					
Amnion					
Allantois					

3. **Twelve-day and eighteen-day embryos:** To open these eggs, crack the large end of the egg with the handle of your scissors and carefully break away the shell with your forceps. Try not to break the *shell membrane* lying below the shell. When you get most of the shell removed, place the egg in a finger bowl filled with saline solution and continue to remove the shell. Below the white shell membrane is another thin, transparent membrane containing many blood vessels. Knowing that the shell is porous, what do you think is the function of this membrane? _____

_____

What has happened to the yolk sac? _____

... to the allantois? _____

After removing these outer membranes, examine the embryo itself. Fill in the information on the chart above.

DISCUSSION:

1. If you could examine the skeleton of the developing chick embryo, you would discover that it develops relatively late. Can you explain this? _____

_____

2. What important structures are among the first to develop, and why? _____

_____

_____

EXERCISE 4-1
PRINCIPLES OF GROUPING AND CLASSIFICATION

INTRODUCTION:

Imagine going into a department store in which there are really no departments. Instead, all the items in the store are displayed at random. It would be very difficult to find a particular item. You might know where some things are, but it would be almost impossible to know where everything is located. So today such stores are departmentalized. Articles with similar uses, or those intended for the same room of the house, are grouped together.

So it is with living things. Biologists have named and identified 1,750,000 organisms. The behavior, activities, and importance of these organisms have been studied. They have also been organized into groups that have similarities, so information about them may found quickly.

But how are things classified? How does a person decide to group certain living things together? This lab activity will give some experience and practice in making those kinds of decisions.

PROBLEM:

To determine how things are grouped or classified.

MATERIALS:

Packets containing sets of construction paper of various sizes, shapes, and color.

PROCEDURE:

A. Classification of Objects

You will be given a package of paper[8] articles that you are to divide into groups. Examine the items closely. List the characteristics of the articles that could be used in their classification. Note that you should begin with characteristics all the pieces have in common and then mention characteristics possessed by only a few or one of them.

---

8.   As an alternative to the paper items, you may collect various inanimate objects that come in a variety of colors, shapes, and sizes.

Fill in the chart below and extend it if necessary to show how you have grouped these articles.

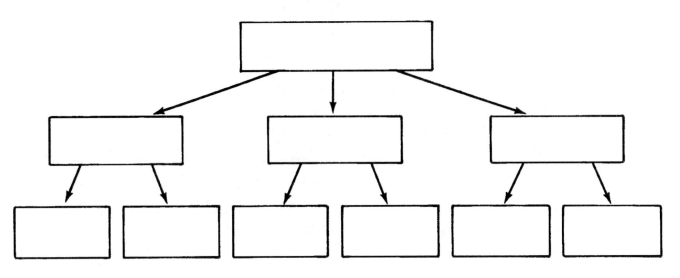

B. Classification of Living Things

Now we will try the same procedure on this set of nine organisms: dog, cat, fish, elephant, bear, deer, frog, snake, kangaroo. What characteristics do these things have? Example: giraffe: has four legs, eats plant material, lives on land, etc.

Make a list for each animal above.

_____

_____

_____

_____

_____

_____

_____

_____

_____

_____

_____

_____

_____

Now construct a chart, similar to the one you did with your pieces of paper, showing how these animals could be grouped.

DISCUSSION:

1. What methods have men used to organize or classify living things? _____

    _____

    _____

2. What do you need to know about an organism before you can classify it? _____

    _____

    _____

3. Are there several ways in which you could have classified the ten animals or the pieces of paper? _____ Which one is the right one? _____

    _____

    Explain. _____

    _____

EXERCISE 5-1
FUNGUS PLANTS:
PLANTS WITHOUT CHLOROPHYLL

INTRODUCTION:

Almost all living things are confronted with the problem of obtaining nutrients. Many plants are green and produce their own food. They are called *autotrophs*, meaning "self-feeders." Other organisms must get their food supply from other plants and animals. These are the *heterotrophs*, meaning "other-feeders." Fungus plants are simple plants that must get their nutrients from organisms or organic matter outside their body. In this activity, we will observe the form and activity of fungus plants.

PROBLEM:

To determine the structure and function of fungus plants.

MATERIALS:

Bread mold: *Aspergillis*; *Penicillium*; petri dishes; culture dishes; yeast; cheese; fruit; wood; grass clippings; Sabouraud's dextrose agar[9]; hand lens; slides; cover slips; small spatula; glycerine; dissecting scope (or hand lens); molasses solution; microscope.

PROCEDURE:

A. Structure of Common Bread Mold

Observe a culture of bread mold with a hand lens or a dissecting microscope. Notice the branching filaments called *hyphae*. There are 2 types of hyphae: Lateral hyphae, called *stolons*, that grow on the surface of the bread; and ascending hyphae, or *sporangiophores* (a stalk or similar structure bearing sporangia), that produce the *sporangium*, which in turn produces spores that give bread mold its characteristic black color. Draw and label a section of this mold showing these structures.

---

9. A culture medium for fungi containing neopeptone or polypeptone agar and glucose, with final pH 5.6; it is the standard, most universally used medium in mycology and is the international reference. Modified Sabouraud's agar (Emmons modification) with less glucose is better for pigment development in the colonies.

B. Culturing Molds

From your teacher obtain cultures of 3 molds: *Rhizopus nigricans*, *Aspergillus niger*,[10] and *Penicillium notatum*. Place a 1 cm square of Sabouraud's agar on the center of 3 slides. Mark these slides "R," "A," and "P." Place a small amount of *Rhizopus* sp. culture on slide "R," *Aspergillus* sp. on slide "A," and *Penicillium* sp. on slide "P." Place the mold along the edge of the agar, and then cover the square of agar with a cover slip. Place these slides in a dark place and observe in 48 hours under low power. To do this, place a drop of glycerine on a clean slide, and remove the cover slip from your culture slide and place it on top of the glycerine. After observing under low power, examine them under high power. You may have to observe these cultures later to see fruiting bodies. From each kind of mold sketch a fruiting body.

C. Variety of Fungus Plants

Examine specimens of other types of molds that you have obtained or your teacher has supplied. Make a sketch of each one. Indicate where it was found, what color it was, what it was growing on, and its relationship to man (beneficial or harmful).

D. Yeast Budding

Yeasts are a slightly different type of fungus. They are one-celled and have no hyphae. They do produce spores, and these are in the dormant stage of the plant in the dry yeast that you buy in the store. Pour a package of yeast into a thin solution of distilled water and molasses. Set this in a warm place at the beginning of the period and come back to examine it later. After at least 30 minutes, place a small drop of the solution on

---

10. *Aspergillus niger* should be handled carefully; if large amounts of the spores are breathed in, they may cause a lung infection known as ***aspergillosis.*** It is less harmful, though not entirely free from risks, if eaten and digested.

a slide and cover it with a cover slip. Examine it under a microscope and notice that many of the small, round yeast cells are bulging at the sides, or have little bumps sticking to them. These are the buds that grow from each cell. When these buds get large enough, they merely break off and form new yeast cells. This is called budding, and is an important method of reproduction in yeast. Draw a diagram of several different stages of budding that you can see under your microscope.

DISCUSSION:

1. List three common features that can be found in all true fungus plants.

   _____

   _____

   _____

2. What are three helpful fungus plants, and how do they benefit man?

   _____

   _____

   _____

3. Name three fungus plants that are harmful and describe how they are dangerous.

   _____

   _____

   _____

4. If bread is left for a period of time in a moist, dark, warm place, it begins to mold. Is this an example of "spontaneous generation"? _____ Explain. _____

   _____

   _____

EXERCISE 5-2
BACTERIA: THE FISSION FUNGI

INTRODUCTION:

Bacteria are some of the smallest organisms you will be able to see in your biology activities. They are so small that you will need a microscope to see them. They are found in all habitats and many are beneficial, whereas others are dangerous to man. In this lab activity we will learn how we can see bacteria and how they can be studied.

PROBLEM:

To learn some techniques used to observe bacteria.

MATERIALS:

Microscope; bean and fruit cultures of bacteria; depression slides; cover slips; methylene blue stain; bunsen or alcohol burners; inoculating loop; vaseline.

PROCEDURE:

A. The Hanging Drop Mount for Looking at Living Bacteria

Place a ring of vaseline around the edge of the depression on the slide. Flame the inoculating loop and transfer a small drop of bean culture with it to the center of a cover slip. Flame the inoculating loop again to sterilize it. Invert this cover slip over the depression on the slide. Observe this slide on low power and then on high power.

What do you observe? _____

_____

Can you see bacteria or chains of them? _____

Are they moving? _____

Draw a few of the bacteria that you see and make another drawing to show how they move.

B. The Fixed and Stained Bacteria Slide

Flame your inoculating loop and put a drop of fluid from a bacterial culture on the center of a clean slide. Spread the drop out very thin. Flame the inoculating loop again and lay it aside. After the fluid has dried on the slide, pass it through the top of the alcohol or Bunsen burner several times. This will cause the bacteria to stick to the slide. Place a drop of methylene blue stain on the slide and allow to stain for 1 or 2 minutes. Rinse off the excess stain by dipping the slide into a beaker of water. Blot dry and observe under a microscope on low and high power. Make a sketch of what you observed on high power.

DISCUSSION:

1. What are some characteristics of bacteria? _____

    _____

    _____

2. Why are bacteria often stained before they are viewed under the microscope? _____

    _____

3. Why should the inoculating needle be flamed before and after use? _____

    _____

EXERCISE 5-3
ALGAE: GRASS OF THE SEA

INTRODUCTION:

Most of us think of trees, flowers, and shrubs when someone mentions the word "plant." Yet the most important green plants may be those that we seldom see. It has been said that plants living in the ocean account for the production of 90% of the oxygen in our atmosphere. In planning space flights involving years of travel time, scientists have considered using algae as a possible source of food and oxygen.

PROBLEM:

To determine the structures and functions of some typical forms of algae.

MATERIALS:

Culture of spirogyra; diatoms; volvox; students' samples of pond water; prepared slides of conjugating spirogyra.

PROCEDURE:

A. **Vegetative *Spirogyra***

Place a few strands of this alga on a slide with a few drops of water from the culture and cover with a cover slip. Locate a filament of this alga on low power and then switch to high power. Make a drawing of the cells that appear in the field of vision on high power. If you have focused carefully, you will notice that a spring-shaped *chloroplast* coils through the cell. On the chloroplast you may find round bodies called *pyrenoids* (pī•rē′noids). These are food storage sites. The rest of the *cytoplasm* of the cell is concentrated near the *cell wall*. Large areas of the cell may appear to be empty. They are really *vacuoles* and are filled with water *cell fluid*. The *nucleus* lies near the center of the cell, imbedded in strands of cytoplasm. In your drawing, label: cell wall, cell membrane, chloroplast, pyrenoid, cytoplasm, nucleus, and vacuole.

B. Conjugating *Spirogyra*

Get from your teacher a prepared slide of conjugating spirogyra. These cells were carrying on a kind of sexual reproduction. Look for two strands of algae that are joined together. During conjugation, the strands of cells grow lateral projections in such a way that adjacent cells are joined, as pictured in the text (Figure 12-5, page 137). This is the *conjugation tube*. Once the filaments of cells are joined, the contents (gametes) of the cells of one strand move through the conjugation tube into the other strand. These two cells' contents fuse, forming a *zygote*. The zygote is a stage that can resist adverse conditions and start growing when better conditions exist. Draw a portion of conjugating spirogyra and label the various stages and parts you can see.

C. Diatoms

Diatoms are the most significant portion of the marine *phytoplankton* (plant plankton). They are unusual in that they possess shells made of silica. As these algae die, their shells accumulate on the ocean floor in the form of diatomaceous earth. This is an important filtering and polishing material used in industry. Prepare a wet mount of a diatom culture and observe all parts of the slide. Draw the various forms of diatoms that you find.

D. *Volvox*

Volvox is a colonial flagellate form of algae. Spherical colonies of this alga are visible to the unaided eye. Prepare and observe a wet mount of *Volvox* on low power. Inside the sphere are dark objects. These are *daughter colonies*. On high power, focus on the edge of a colony to view the cell layer. Draw a single *Volvox* colony as seen on lower power.

E. Samples of Pond Water

Prepare wet mounts of the pond water samples you collected. Include drawings of the forms you find.

DISCUSSION:

1. In what ways (other than those mentioned above) might algae be useful to man?

_____

_____

_____

_____

2. The practice of referring to algae as plants has been criticized.

    What possible reasons can you give for this? _____

_____

EXERCISE 5-4
THE AMEBA, PARAMECIUM, AND EUGLENA

INTRODUCTION:

Can something almost invisible to the unaided eye be alive? For many years before the development of the microscope, lists of living things included only those things in plain sight. Hardly anyone imagined that a whole world of invisible living, breathing, reproducing organisms existed. The thought of thousands of these tiny organisms living in a single drop of water stirs the imagination like the best of science fiction stories; and yet, they really exist. In this activity we will take a look at a few of the creatures we find in this fascinating microscopic world.

PROBLEM:

To examine the variety of structure and function in the protozoa, and to observe their life processes.

MATERIALS:

Cultures of *Ameba proteus, Paramecium multinucleatum, Euglena viridis*; depression slides; cover slips; methyl cellulose; cotton fibers; salt; toothpicks; distilled water.

PROCEDURES:

A. *Ameba proteus*

Obtain a drop of water from the bottom of an ameba culture. Place this in the center of a depression slide and cover with a cover slip. Mount this slide on your microscope and with low power search the drop of water for an ameba. Since these cells are transparent, they are almost invisible except under dim light, so you may have to adjust the microscope diaphragm carefully to be able to see them. When you have located an active ameba, observe it for 5 minutes, making a sketch of its appearance at the end of each minute.

How does the ameba move? _____

What generally can be said about an ameba's shape? _____

_____

Now place a crystal of salt into the drop of ameba culture. What kind of reaction do you see? _____

B. *Paramecium multinucleatum*

On a clean slide place a few strands of cotton. Place a drop of paramecium culture on top of these fibers and cover with a cover slip. Observe under low power. How do these organisms respond when they contact the fibers? _____

Compare the shape of the paramecium to the ameba. _____

Locate the *nucleus* in the center of the paramecium. Switch to high power and examine the edge of the paramecium. Notice the *cilia* and their wave-like action. Locate also the *oral groove* and *gullet*. Watch for the formation of a *food vacuole* at the base of the gullet and observe its movement. Look at one end of the paramecium and locate a clear, bubblelike structure that seems to appear and disappear repeatedly. This is the *contractile vacuole*. It pumps excess water out of the paramecium. How often does this vacuole contract in a minute? _____

Place a few drops of distilled water next to the cover slip and draw it into the slide by blotting up water at the opposite end. Explain the change that occurs in the activity of the contractile vacuole. _____

_____

_____

Draw a paramecium as seen on high power and label all the organelles that you could identify. (An organelle is any structure within a cell.)

C. *Euglena viridis*

Place a drop of methyl cellulose in the center of a slide. Add a drop of euglena culture and mix with a toothpick. Observe on low power. Euglena appear to move in two ways. One is by a *flagellum*, or whiplike structure, that acts like a propeller. Another is by that alternate contraction and extension of its body called the euglenoid movement. Center your slide on one euglena and switch to high power to locate its organelles. Notice the green structures in the cytoplasm. What is their function? _____

_____

Near the base of the flagellum is a pigmented structure called an eyespot. What is its value to the euglena? _____

_____

Draw your euglena as it appears on high power and label: flagellum, nucleus, eyespot, chloroplasts.

EXERCISE 6-1
THE HYDRA AND PLANARIA

INTRODUCTION:

Years ago there was a dramatic report of a boy who had been hitching a ride on a freight train and as the train went through a tunnel, the boy's arm had been severed at the shoulder. Quick medical work reattached the arm, and the boy has some use of that arm today. Another man recently had his thumb, which had been severed and lost, replaced by one of his great toes. Man must rely on his intelligence to replace missing parts.

The crayfish or "crawdad" is known to actually remove an appendage that has been damaged. Then a new one grows to take its place. The organisms we will be studying in this activity also have this regenerative power.

PROBLEM:

To determine how the hydra and the planaria carry out their life functions and to observe how they regenerate missing parts.

MATERIALS:

Hydra and planaria cultures; slides; petri dishes; scalpels or razor blades; daphnia culture; liver or hard-boiled egg yolk; hand lens or dissecting scope; dissecting needle; medicine dropper; vaseline; depression slides; epsom salts; black paper; microscope.

PROCEDURE:

A. **Hydra**

Place a few drops of water on a slide and place a hydra on the slide. Using a piece of white paper as a background, view with a hand lens or the dissecting scope. Observe carefully without disturbing the slide for one minute. Place a daphnia on the slide with the hydra and watch for a few minutes. While viewing the hydra through the hand lens, touch one of its *tentacles* with the tip of a dissecting needle. How does the hydra respond? _____

Cover the slide with a cover slip and observe under low power. Locate the *tentacles, mouth,* and *basal disk.* Locate specialized stinging cells called *nematocysts.* Along the main body region, locate the outer layer of cells called the *ectoderm* and an inner layer called *endoderm.* Between these layers you may find the nerve net. Also locate the central cavity, called the *gastrovascular cavity.* Draw the hydra and label: tentacles, mouth, gastrovascular cavity, ectoderm, endoderm, nematocysts, nerve net.

Can you locate any "buds" on your hydra? _____

Describe them. _____

B. Hydra Regeneration

Place some well-fed hydra into fresh water and with a scalpel cut one laterally in half between the mouth and the basal disk. Cut another hydra in half longitudinally. Place each of these sections into a separate petri dish and observe for a week. Keep a record of your observations.

Place another hydra on a cover slip in a drop of pond water and, with a slide placed over the cover slip, mash the hydra. Remove the slide. Add one or two drops of water to the cover slip and mount on a depression slide. Seat with vaseline. Observe each day for a week. Keep a record of your observations.

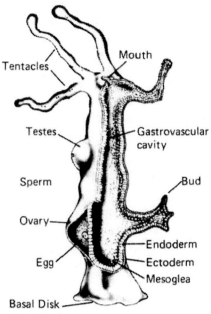

C. The Planaria

Place a hungry planaria in a petri dish with a small amount of aged water. With a hand lens, observe the planaria's activity. Does it seem to have a head end? _____

Explain. _____

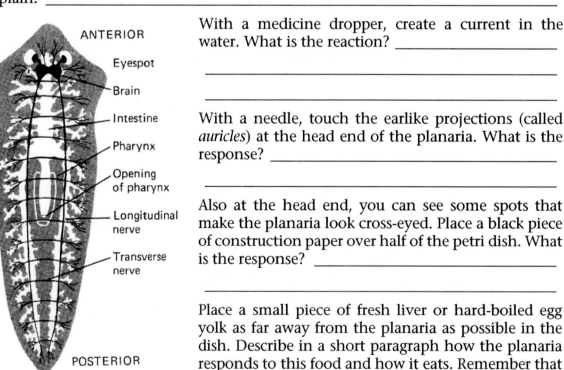

With a medicine dropper, create a current in the water. What is the reaction? _____

_____

_____

With a needle, touch the earlike projections (called *auricles*) at the head end of the planaria. What is the response? _____

_____

Also at the head end, you can see some spots that make the planaria look cross-eyed. Place a black piece of construction paper over half of the petri dish. What is the response? _____

_____

Place a small piece of fresh liver or hard-boiled egg yolk as far away from the planaria as possible in the dish. Describe in a short paragraph how the planaria responds to this food and how it eats. Remember that the mouth is located on the ventral side of its body.

_____

_____

_____

_____

_____

Observe a prepared slide of the cross section of a planaria on low power and make a drawing of this. Label: ectoderm, endoderm, mesoderm, cilia.

D. Planaria Regeneration

Obtain a live planaria that has not been recently fed. Shown at the right are some of the possible cuts that can be made. Lab teams should select different cuts to carry out. Place a few crystals of epsom salts in the petri dish to relax the planaria. Place the planaria on a slide with a few drops of water. Use a dissecting scope to view your specimen while you make your desired cuts with a clean, sharp scalpel or razor blade. Be sure to cut all the way through the body. Place each piece of the planaria in a separate petri dish with clean, aged water and put in a cool, dark place. Observe daily for a week and keep a record of your observations with sketches of the changes detected on each day.

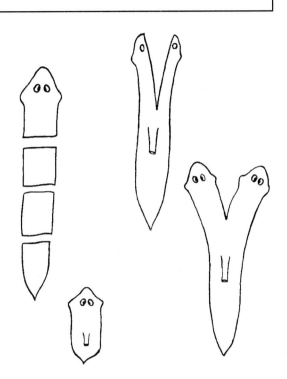

EXERCISE 6-2
THE ROUNDWORMS

Phylum Nematoda

INTRODUCTION:

Although many people are unaware of them, roundworms are very abundant on the earth. Roundworms may be found in the soil and in fresh or salt water. Many of the roundworms are parasitic, living in the bodies of plants, animals, or man. It has been said that as much as one fourth of the world's entire human population is infected with some sort of parasitic roundworm (genus *Ascaris*), this figure being much higher in some countries of the world.

PROBLEM:

To become familiar with some of the common roundworms that may be parasitic to man and to develop skills of dissection.

MATERIALS:

Prepared slides of hookworms and trichina worms; microscope; preserved specimens of *Ascaris suum* (or *A. lumbricoides*); dissection pan; pins; dissection tools (*see kit below*).

PROCEDURE:

A. **Microscopic Roundworms**

Examine under the microscope prepared slides of the hookworm, the trichina worm, and any other roundworms available. Identify as many of the structures as you can (refer to page 166 of the text). Refer to your textbook for information concerning the life cycle of these worms. What precautions could prevent infection by the hookworm?

_____

... by the trichina worm? _____

_____

B. **Dissection Instructions**

1. Prepare your dissection kit.

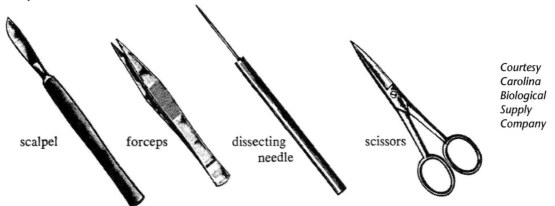

scalpel   forceps   dissecting needle   scissors

Courtesy Carolina Biological Supply Company

2. Become familiar with these terms:

    anterior   posterior   dorsal

    ventral   longitudinal   cross section

    median

3. Read all directions before starting to cut anything, and be sure you understand exactly what you are to do.

4. For best control, hold your specimen with your fingers rather than with forceps. The forceps are for lifting small objects, or for using as a probe, if the points are not sharp.

5. Work carefully and neatly; dissect, do not "butcher."

C. External Anatomy of the *Ascaris*

Lay your specimen in the dissection pan and examine its external features. If you have a male, its posterior end will be strongly hooked. This region is used as a reproductive organ in transferring sperm to the female's body. At the end of the hook is the *cloacal opening* from which sperm are passed out and into the *genital opening* of the female, which is about one third of the way back from the front, and on the ventral surface. Be careful not to confuse the female's genital opening with the *excretory pore* which is just behind the mouth in both sexes. The *mouth* itself is composed of three fleshy *lips* and is on the extreme anterior end of the body. At the posterior end of the body is the *anus*, which in males is the same as the cloacal opening. You should also be able to see a *dorsal line* and a *ventral line* as well as two *lateral lines* extending the length of the body. The entire body is covered with a tough cuticle which helps to protect the body.

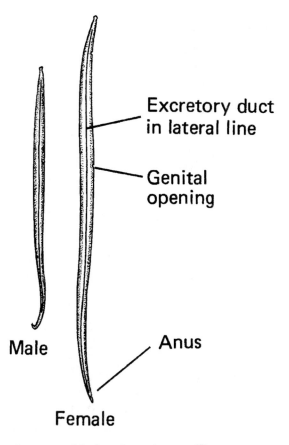

As an intestinal parasite, what sort of protection would the *Ascaris* need? _____

_____

_____

_____

D. Internal Anatomy

Now take your scalpel and make a short cut through the mid-dorsal body wall near the posterior end of the *Ascaris*. Be careful not to cut too deeply, or you will damage the internal organs. Place one end of your scissors into this opening and cut forward along the entire length of the body, being careful not to point the scissors downward and rip the organs apart. Once this cut is made, pin back the body walls so that you can observe the internal organs.

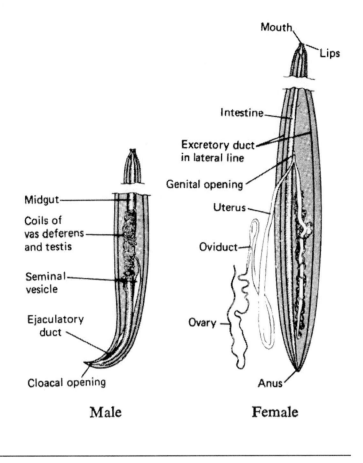

Compare your specimen to the drawings of the *Ascaris* in the figure at the right. Locate the esophagus and the long intestine ending at the anus. What types of organs that are normally found in digestive systems of animals are missing? _____

Why? _____

The *ovaries* of the female are a mass of small, coiled tubes, which lead into the larger tubes of the *oviduct* and to the larger and straighter *uterus*. The paired uteri join at the *vagina* where they are connected to the outside of the body by the genital opening. The *testes* of the male are also a mass of coiled tubes which lead into the slightly larger *vas deferens* to the much larger *seminal vesicle* near the posterior end of the body. Sperm pass from the seminal vesicle through the *ejaculatory duct* and out of the cloacal opening which (in males) is the same as the anus.

The *lateral lines* contain *excretory canals* which lead forward to the excretory pore near the mouth. These function in ridding the body of waste in a manner similar to that of the human kidneys.

With a great deal of skill and careful work, you may be able to locate the ring of *ganglia* which forms a branching mass around the esophagus. This is the center of the simple nervous system of the *Ascaris*. Do you find any sense organs? _____

There is no distinct respiratory or circulatory system in the *Ascaris*. Transportation of materials is accomplished by osmosis from cell to cell, and by movements of fluids throughout the body cavity. Most of the respiration is *anaerobic* (without oxygen) since there is little free oxygen available within the intestine of the host.

DISCUSSION:

List as many ways as possible in which the *Ascaris* is adapted to a parasitic way of life.

_____

_____

_____

_____

_____

_____

EXERCISE 6-3
DISSECTION OF THE EARTHWORM

Phylum Annelida: Segmented Worms

INTRODUCTION:

Annelids are different from all other worms in that their bodies are divided into many segments. In many annelids, these segments are visible from the outside, but in others they can be seen only after dissection. Within the group are marine, aquatic, and soil-living forms of animals, which vary from as small as one millimeter to over eleven feet long. Most commonly known is the earthworm or "night crawler" of the genus *Lumbricus*.

PROBLEM:

To become familiar with the anatomy of the common earthworm, and to note differences from the anatomy of other organisms.

MATERIALS:

Earthworm; dissection pan; pins; dissection tools; microscope; prepared slide of earthworm cross section; hand lens.

PROCEDURE:

A. **External Anatomy**

Compare your worm with Figure 1. What structure can be used to identify the anterior end? _____

... the dorsal side? _____

Rub your finger over the ventral surface and lateral surfaces and note the bristlelike setae, which are found on each segment. These provide the earthworm with a firm grip of the soil, as you know if you have ever tried to pull one out of the ground. Locate the anus and the mouth. Note that above the mouth is a bulging prostomium; can you guess its purpose? _____

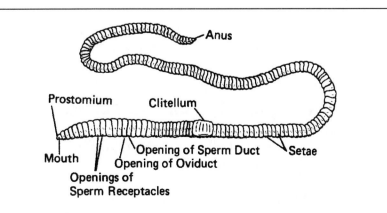

EXTERNAL ANATOMY OF THE EARTHWORM

65

Start counting the segments from the mouth backwards. Locate the ventral *opening* of the *seminal receptacles* between segments 9 to 11. Sperm enter the worm and are stored in these seminal receptacles until they are needed later. On segment 14 are the *openings* of the paired *oviducts*, while segment 15 contains the *openings* of the paired sperm ducts (*vas deferens*).

The earthworm is *hermaphroditic*, containing reproductive organs of both sexes; therefore, every earthworm has all of the structures listed above. Although each earthworm produces both sperm and eggs, it cannot fertilize its own eggs but must mate with another worm. Two worms join their anterior segments (*see figure below*) by secreting *slime rings* from the smooth *clitellum* (segments 32 to 37) that surround both worms. Sperm are released from each seminal vesicle into this slime ring, and as the two worms separate, the sperm travel through the slime ring to the openings of the other worm's seminal receptacle, where the sperm are stored. Later, when the eggs mature, the sperm fertilize them inside another slime ring, which slides off of the body to form the cocoon in which the eggs will develop. Note that in mating, each worm receives sperm from the other one.

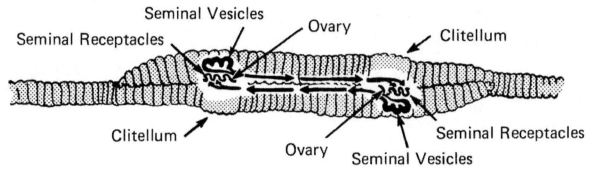

REPRODUCTION IN EARTHWORMS

B. Internal Anatomy

Lay the earthworm in your dissection pan with the dorsal surface up. Start cutting through the body wall on a mid-dorsal line at least one inch behind the clitellum. Since the important organs are anterior to the clitellum, starting your cut behind it will give you practice in avoiding the internal organs before you reach a critical area. Continue along this cut on the mid-dorsal line up to the region of the mouth. Once the cut is finished, try to fold back the skin and pin it down as you did with the *Ascaris*. You will notice that between each segment is a wall of connective tissue, the *septum*. Break these septa with your probe or needle, so that the skin can be folded back to reveal the internal organs. The skin is covered with moist secretions, which enable gases to be absorbed for respiration. Under the skin are two layers of muscles; *longitudinal muscles* that shorten the worm when contracted, and *circular muscles* that extend the worm when contracted. These work with the setae to provide movement.

The digestive system consists of the *mouth cavity*, the thick *pharynx*, followed by the *esophagus* (segments 6 to 15), a large *crop* for storage, the slightly larger and muscular *gizzard* that grinds the food, the long *intestine*, and the *anus*. What does the earthworm eat, and how does it get its food? (Clue: cut open a section of the intestine behind the clitellum.) _____

Surrounding the esophagus are 4 rings called the *aortic arches* (hearts) connected to a *dorsal* and a *ventral blood vessel*. Near the aortic arches you will find the large white seminal vesicles (segments 10 to 12) that store the sperm (male function) and, near them, the smaller white *seminal receptacles* that are used for storing the sperm (female function) from another worm after mating.

Along the entire body, in each segment, are the very small, coiled tubes called the *nephridia*. To see these best, cut out a section of the intestine posterior to the clitellum and search the body walls. A hand lens may be useful. The nephridia are excretory organs.

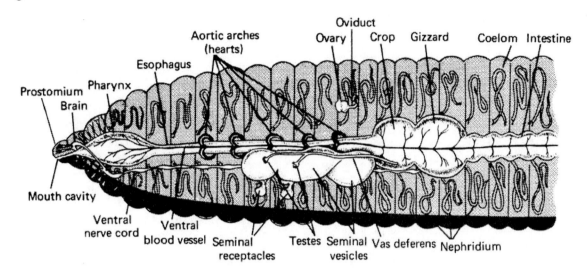

INTERNAL ANATOMY OF EARTHWORM

Surrounding the pharynx are two nerve fibers that join dorsally to form a *ganglion* (suprepharyngeal ganglion) and they join again ventrally to form another ganglion (subpharyngeal ganglion). These ganglia are sometimes referred to as the "brain," but are in fact much simpler than a true brain. From the ventral ganglia the *ventral nerve cord* extends the length of the body. You can usually detect a small ganglion on each segment of this nerve.

Examine a prepared slide of a cross section from an earthworm under the microscope. Locate the dorsal and ventral blood vessels, the ventral nerve cord, nephridia, and the two different layers of muscles. Examine the intestine and notice that the inside of it is not round, but contains a loop called the *typhlosole*. What advantage, if any, might this give to the digestive system of the earthworm? _____

_____

DISCUSSION:

1. The earthworm has both male and female sex organs, yet cannot fertilize its own eggs. What advantage (if any) is there in such "cross-fertilization"? _____

_____

2. Since the formation and efficient functioning of the slime ring is essential for reproduction, does it seem likely that the ability to produce it occurred by slow evolutionary development over many generations? _____ If so, how was the species propagated during the early stages of the history of the species? _____

   _____

3. Compare the digestive system of the earthworm with that of the *Ascaris* and explain why the differences exist and are necessary. _____

   _____

   _____

   _____

EXERCISE 6-4
DISSECTION OF THE CLAM

Phylum Mollusca
Class Pelecypoda: Hatchet Foot

INTRODUCTION:

Members of the class Pelecypoda provide delight for epicureans, jewelers, and artisans the world over, because they provide food, pearls, and mother-of-pearl, which can be fashioned into hundreds of forms. Another name for this group is "bivalves," as they possess two shells or *valves*. Included in the group are clams, oysters, mussels, scallops, and shipworms. They vary in size from one-half inch across up to four feet wide and can weigh more than 500 pounds (the giant clam of the South Pacific). Since clams and mussels are found both in salt and fresh water, they are common throughout the United States and the entire world.

PROBLEM:

To become familiar with the anatomy of the clam, and to note differences from the anatomy of other organisms.

MATERIALS:

Clam; dissecting pan; dissecting tools.

PROCEDURE:

A. **External Anatomy**

Examine your clam and notice that it has two *valves* (or shells) that are joined at the *hinge*. Although the clam has no head, it does have a "front and back" (*anterior* and *posterior*, respectively). The edge along the hinge is the *dorsal side*, while the open edge is the *ventral side*. The anterior end is the one closest to the bulge, or *umbo*. This umbo is the oldest part of the clam, and you can see the concentric *growth rings* extending outward from it.

The shell is composed of three layers: the outer *horny layer*—very tough, dark, and uneven; the middle *crystalline layer*; and the inner *pearly layer*—the part from which "mother-of-pearl" comes.

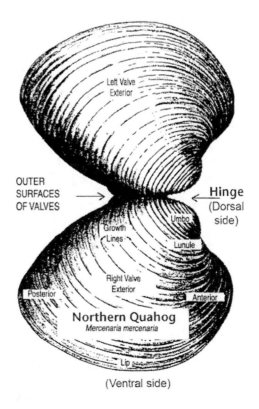

(Ventral side)

69

B. Internal Anatomy

If you use the proper technique, clams are easy to open. Live clams open if heated, but the easiest method for opening most clams is to slip your scalpel blade between the valves and cut the two large *adductor muscles* near the hinge. Check the figure for the exact location of these muscles and be careful not to cut too deeply.

As you open the valves, you will see the thin membrane of the *mantle* which is attached to the valves. The mantle secretes substances that produce the valves, and it also produces pearls. Pearls are formed when a particle of sand gets between the mantle and the valves and irritates the clam. In response, the clam secretes from the mantle successive layers of pearl around the particle. The smooth, round pearl that is the final product does not irritate the soft membranes of the clam. The only highly valuable pearls come from marine pearl oysters, but many clams and mussels do produce less valuable pearls, and you might find one as you dissect your clam.

At the posterior end of the mantle, two groovelike passageways may be found: the dorsal *excurrent siphon* and the ventral *incurrent siphon*. Water currents enter the incurrent siphon and leave by the excurrent siphon.

Position your clam with the ventral surface nearest you, the anterior end to your left (*see figure at the right*). Remove the mantle from the side that is uppermost to reveal the *visceral mass*. This includes the entire clam, except the shell, mantle, gills, and *foot*, which is ventral and somewhat anterior to the visceral mass. The foot is somewhat hatchet-shaped (hence the name of the class) and is a muscular organ

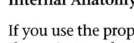

used for locomotion along the bottom. Covering much of the visceral mass are the sheetlike *gills*, which have many lines on them and are filled with blood vessels to pick up oxygen from the water. If you have a female, you might find small bumps in the gills which are the larva of the clam. Ordinarily these would soon leave the female's body and attach themselves to the fins or gills of fish until they are mature.

The gills are covered with a mucus lining and countless small cilia. Small food organisms coming into the clam through the incurrent siphon are trapped in this mucus and then swept toward the two flaplike *labial palps* that surround the mouth.

Now remove the foot and visceral mass from the shell by gripping the entire unit firmly and pulling it out. Directly under the hinge area is the *heart*, located inside of a dark pericardial sac. The heart is composed of a *ventricle* and two *auricles*. The heart pumps blood through an *anterior aorta* and a *posterior aorta* to the rest of the body. Posterior to the heart are the dark *kidneys*, which function in the excretion of wastes.

Take the foot and the visceral mass and cut them in half, starting on the ventral side of the foot, passing through it along the flat axis, and continuing dorsally through the visceral mass. This will leave two flat platelike sections, revealing the digestive tract as shown at the right. The short *esophagus* leads from the mouth to the *stomach*, which is surrounded by the dark *digestive gland*. The *intestine* will appear as a mass of coiled tubes (some cut open, revealing their irregular inner surface). Surrounding the intestines are the granular *gonads*, which may be either ovaries or testes, but this distinction is indiscernible without a microscope. The intestine passes the heart and empties through the anus near the excurrent siphon.

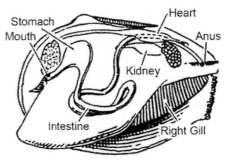

DISCUSSION:

1. Using an encyclopedia or Internet source, look up the phylum Mollusca. What other types of animals are in this phylum? _____

   _____

2. What characteristics of the clam are also found in these other mollusks? _____

   _____

3. What is meant by "cultured pearls," and how are they formed? _____

   _____

   _____

4. List as many similarities and differences as you can between the mollusks and the annelids. _____

   _____

   _____

EXERCISE 6-5
DISSECTION OF THE STARFISH

Phylum Echinodermata

INTRODUCTION:

The starfish is a member of the phylum Echinodermata, which means "spiny skin." The name fits well, as you can readily see from the many spines on its surface. Other members of the phylum, such as the sea urchin, have much larger and sharper spines.

PROBLEM:

To discover the structure and functions of the starfish, and to note differences between the starfish and other organisms.

MATERIALS:

Preserved starfish; dissection tools; dissection pan.

PROCEDURE:

A. Examination of a Starfish

Lay the starfish in your dissection pan upside down. The *oral surface* (ventral) is named for the circular *mouth* in its center. Note that the *oral spines* are somewhat longer than the other spines. Along the center of the ventral surface of each *ray* (arm) is the *ambulacral* (ăm'byə•lă'krəl) *groove*,[11] which contains the soft, extendable *tube feet*. At the very tip of each ray is a small *tentacle* and a pigmented light-sensitive *ocellus* (ō•sĕ'ləs, eyespot). In between the spines and protected by them are soft *gills*, which may require magnification to see (*see figure at the right*).

Now turn your starfish over, onto its aboral (*ab*, "away from"; *oral*, "mouth") or dorsal surface. Just off center is the hard, round *madreporite* (mă'drə•pōr•ĭt).[12] Two rays are nearer to the madreporite than the rest; these are called the *bivium* (bĭ'vē•əm). The other three rays are called the *trivium* (trĭ'vē•əm). In the very

Anus, Madreporite, Pyloric caecum, Gill, Tentacle, Eye spot, Spine, Ossicles, Gonad, Ambulacral groove, Tube feet

---

11. *Ambulacral* refers to the radial areas of echinoderms; the principal nerves, blood vessels, and elements of the water-vascular system run along these radial areas.

12. *Madreporite* is a perforated or porous body located at the distal end of the stone canal in echinoderms

center of the aboral surface is the *anus*, which may be hard to see. Use a blunt probe to locate it.

B. Dissection of a Starfish

Make a cross-sectional cut with your scissors about halfway from the tip of one of the trivium. On the top half of the inside you will find the large glandular *gastric caeca* (sē′kə, *sing.* caecum; also, *pl.* ceca, *sing.* cecum) or digestive glands. Often these almost fill the entire ray. Describe them. _____

_____

Below these you will find the *gonads* if your starfish is in reproductive condition. These reproductive organs appear identical in both sexes. Describe their appearance. _____

_____

Now take one of the other rays of the trivium and make a V-shaped cut along the dorsal surface. Cut with the apex of the V at the tip of the ray and extend your cuts back to the central disk along the edges of the ray, so that a top layer can be folded back and cut off, exposing the internal organs. Remove the gastric caeca and the gonads. Running down the internal ventral surface is the *ambulacral* (ăm′byə•lă′krəl) *ridge* which is opposite the ambulacral groove that you examined earlier. On the inside of this ridge is the *radial canal*. Along the sides of the ridge are soft, bulbular *ampullae* (ăm•pŭ′lə).[13] These are attached at one end to the tube feet, and at the other end to the radial canal. You may be able also to find the two *retractor muscles*, which are attached to the stomach. What is their function? _____

_____

Carefully cut around the top of the central disk, starting by one of the bivium and Moving away from the madreporite (mă′drə•pōr•īt) around the second bivium. This will leave the section between the two rays of the bivium (including the madreporite) uncut. Carefully fold this cut area back to expose the organs of the central disk. If you work carefully, you may still find where the *intestine* leaves the *stomach* and travels a short distance to the anus. The large, saclike stomach is 5-lobed and covers much of the central disk.

Look below the madreporite and find the *stone canal* connected to it. This moves downward to a *ring canal* which surrounds the mouth region. These structures then connect to the radial canals, forming what is called the water-vascular system. (*See figure at the right.*) Water can flow from the tube feet through the radial canals to the ring canal, then up the stone canal and out of the madreporite.

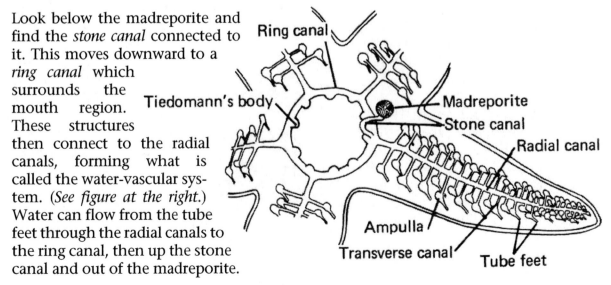

13. An *ampulla* is a saclike anatomic dilation (i.e., swelling) or pouch.

DISCUSSION:

1. How is the water-vascular system used by the starfish? _____

    _____

    _____

    _____

2. Describe the eating habits of a starfish and explain how its anatomy makes these habits

    possible. _____

    _____

    _____

    _____

    _____

3. Can you name any other animal that might use a system similar to the water-vascular

    system of the starfish? _____

4   Does this indicate an evolutionary relationship? (Be sure to include pro and con points

    of view in your answer.) _____

    _____

    _____

    _____

    _____

EXERCISE 6-6
DISSECTION OF THE CRAYFISH

Phylum Arthropoda
Class Crustacea

INTRODUCTION:

Crayfish ("crawdads" or "crawfish," as they are sometimes called) are freshwater crustaceans that are very similar to lobsters. You can readily find these in freshwater ponds and streams in much of the United States. Watch for their chimneylike mud mounds in the shallow waters. If you try to catch them, you will soon discover their amazing ability to make a hasty retreat—backwards!

PROBLEM:

To discover the structure and functions of the crayfish, and to note differences between the crayfish and other organisms.

MATERIALS:

Preserved crayfish; dissection tools; wax dissection pan.

PROCEDURE:

A. **Examination of a Crayfish**

Examine the external features of your specimen carefully and compare them with the figure below. Crayfish have a hard exoskeleton made of chitin. In what two ways does this exoskeleton function? _____

In crayfish, the exoskeleton is divided into two parts: the *cephalothorax* (the anterior half) and the *abdomen*. How many segments does the abdomen have? (The telson is one segment.) _____

At the front of the cephalothorax is the *rostrum*, which projects out over the *compound eyes* like a beak. What is unusual about the placement of the eyes? _____

_____

The rear portion of the cephalothorax is the *carapace*, which is a single, shieldlike piece covering the sides and dorsal surfaces.

Courtesy Carolina Biological Supply Company

77

An interesting thing about the crayfish is the great number and variety of appendages that it has. You should remove one of each kind of appendage, proceeding down the right side of the body. Starting with the most anterior appendages, remove one *antennule* and one *antenna*. (*See figure at the right.*) Be careful to remove the entire appendage (they are branched) by twisting them off at the base with your forceps.

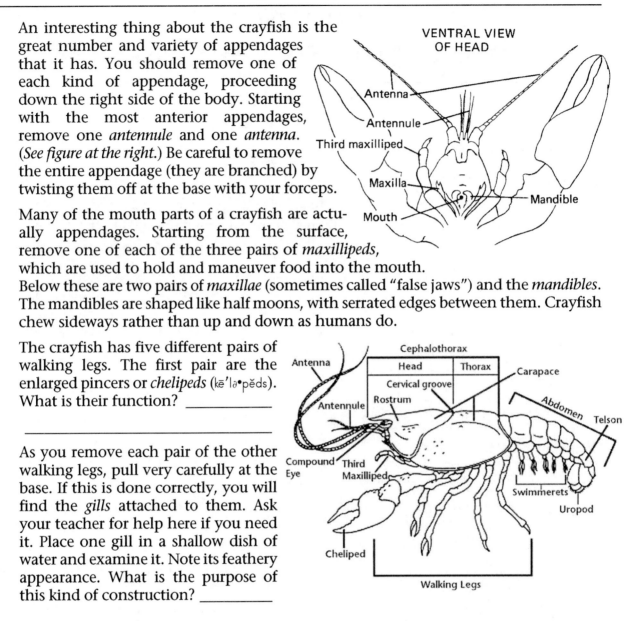

VENTRAL VIEW OF HEAD

Antenna
Antennule
Third maxilliped
Maxilla
Mouth
Mandible

Many of the mouth parts of a crayfish are actually appendages. Starting from the surface, remove one of each of the three pairs of *maxillipeds*, which are used to hold and maneuver food into the mouth. Below these are two pairs of *maxillae* (sometimes called "false jaws") and the *mandibles*. The mandibles are shaped like half moons, with serrated edges between them. Crayfish chew sideways rather than up and down as humans do.

The crayfish has five different pairs of walking legs. The first pair are the enlarged pincers or *chelipeds* (kē′lə•pĕds). What is their function? _____

Antenna
Cephalothorax
Head   Thorax
Cervical groove
Carapace
Rostrum
Antennule
Abdomen
Telson
Compound Eye
Third Maxilliped
Swimmerets
Uropod
Cheliped
Walking Legs

_____

As you remove each pair of the other walking legs, pull very carefully at the base. If this is done correctly, you will find the *gills* attached to them. Ask your teacher for help here if you need it. Place one gill in a shallow dish of water and examine it. Note its feathery appearance. What is the purpose of this kind of construction? _____

_____

How do the other walking legs differ from each other? _____

_____

On the abdomen, you will find five pairs of *swimmerets* if you look closely. In males, the first pair (and to a lesser degree the second) is greatly enlarged. These are used to transfer sperm to the female. In females, all swimmerets are approximately equal in size. Females will also have a large oviduct opening at the anterior of the abdomen. What is the sex of your specimen? _____

The last pair of appendages are the flattened *uropods* (yŭr′ə•pŏds). These are attached to the telson, and the three form a flipper which is used in rapid backward swimming.

B. **Dissection of a Crayfish**

Take your scissors and carefully remove the carapace by starting to cut upward just below the rostrum and continuing up around the sides and back. Be careful not to cut too deeply so that you will not destroy the internal organs. Gently loosen the rest of

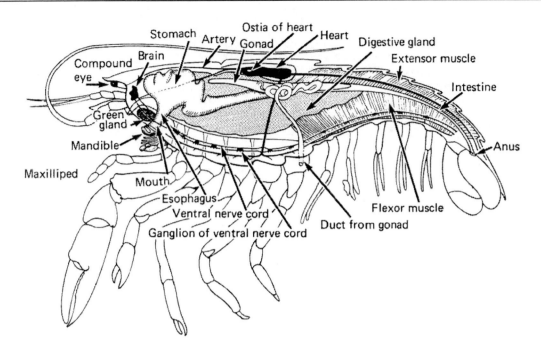

the carapace and lift it off. If you have done this expertly, you will find the *heart* in a *pericardial sinus* on the mid-dorsal area. (*See figure above.*) If you do not see it, check the inside of your carapace, to which the heart sometimes adheres. Note the three small openings on the heart, called *ostia*, through which blood enters the heart. Just below the heart you should find the *ovaries*, which are usually large structures with two lobes, or the white *testes* with two coiled sperm ducts attached to them. Beneath these is the *intestine*, which can be traced forward to the large *stomach.* Cut the stomach open and find the hard teeth inside it. How many teeth are there? _____

Under the intestine is the large brown or green *digestive gland* (liver). Anterior to and below the stomach is the *green gland*, which functions like a kidney. Above this, find the *brain*. Two nerve cords lead backward from the brain, form a loop around the esophagus, and unite to form a single *ventral nerve cord*. Look for the bumplike *ganglia* at intervals along this.

DISCUSSION:

1. In what ways is a crayfish similar to an earthworm? _____

_____

_____

2. Do these similarities show an evolutionary link? Discuss the pro and con aspects of such a possibility. _____

_____

_____

_____

_____

EXERCISE 6-7
DISSECTION OF THE PERCH

INTRODUCTION:

Fish come in almost every size and shape imaginable. They vary from the 10-mm long Philippine goby[14] to records of Russian sturgeon[15] over 28 feet long and weighing 3,000 pounds (1,360 kilograms). Obviously, dissecting only one fish will not tell us much about the great many different kinds, but the perch (families: Percidae, Centrarchidae, and Serranidae) does portray the basic characteristics and structures that are found in most bony fish. Its dissection will introduce us to these characteristics and structures.

PROBLEM:

To dissect the perch and learn from it the basic anatomy of all bony fish, and to note differences from other organisms.

MATERIALS:

Preserved perch; dissection pan and tools; slides and cover slips; microscope.

PROCEDURE:

A. External Anatomy

Lay your fish on its side in your dissection pan. Locate the *mouth, nostrils,* and *eyes.* Just behind the mouth and eyes is the flap-like *operculum* (ō•pər'kyə•ləm), which covers the gills. Take your probe and insert it into the operculum and through the gills underneath. Now open the mouth and observe the probe and its path. Now insert another probe into the nostrils. Do they connect with the throat and gills? _____ What is the function of the nostrils? _____

Cut off the operculum on one side of your fish to reveal the gills underneath. A *gill* consists of a double row of feathery *filaments* attached to a cartilaginous *gill arch.* Remove one gill by cutting off the arch at the upper and lower attachments. Try to locate the place where blood vessels enter the gill arch. What is the function of these blood vessels? _____

_____

If you examine the gill, you will notice that anterior to each pair of filaments is a pair of short, comblike *gill rakers,* which protect the gills by blocking entry of food and other hard objects into the gills.

Just posterior to the operculum are the paired *pectoral fins* and the paired *pelvic fins* below them. Along the back of the perch are the *dorsal fins.* How are they supported?

_____

---

14. Gobies are small, spiny-finned fishes of the family Gobiidae, having the pelvic fins united to form a ventral sucking disk.

15. Sturgeons are large, plain-featured fishes found in the Northern Hemisphere. Many species are restricted to freshwater, but several migrate up rivers from the sea to breed in fresh water. Sturgeon use their protruding mouths to grab fishes and invertebrates along the silty bottom. The most significant commercial use of sturgeon is the harvest of their eggs for caviar. There are 24 species recognized in this family.

Just behind the *anus* is the small *anal fin*, and at the very posterior end is the *caudal fin*. Observe fish in an aquarium, if possible. Which of these fins are used for fast moving?

_____

_____

Which are used for slow moving? _____

Along the side of the fish's body, from the operculum to the tail, is a light line called the *lateral line*. This line is composed of a series of small pores which are connected by a canal. Water enters this canal in such a way that change in water pressure, or movement in the water, stimulates nerves in the canal. Its function, then, is one of a sense of touch for the fish.

Remove several scales from your fish and examine them under your microscope on low power. Note the toothed posterior edge, and also the concentric rings on the central portion of the scale. These rings are *growth rings*. They grow larger in spring and summer when food is plentiful, and smaller in fall and winter. Count the number of annual growth rings. How old is your fish? _____ How are the scales arranged on the fish? _____

_____

Draw a picture of one of these scales.

B. Internal Anatomy

Insert the point of your scissors through the body wall of the fish just anterior to the anus. Cut forward along a mid-ventral line until you reach the gills, being careful not to cut too deeply. Now cut dorsally along the edge of the operculum until you reach the back. Make another cut dorsally from your beginning cut in front of the anus. Lift this flap of skin and flesh from the side of the fish and cut it off along the back with your scalpel. You should now have a clear view of the internal organs.

One of the most noticeable organs is the reddish-brown *liver* near the anterior of the body cavity. Attached under this is the saclike *gall bladder*. What color is it? _____ Carefully remove the liver by cutting it from its attachment and find the short esophagus, followed by the curved *stomach*. Near where the *intestine* joins the stomach are the fingerlike *pyloric caeca*, which aid digestion. Follow the intestine back to where it leaves the body at the anus. It may be covered with long masses of fat. Cut away the esophagus from the mouth and the intestine from the anus, removing the digestive tract.

Above the place where the intestines were located are the gonads. If you have a female, you will find the large, yellowish *ovary* containing numerous eggs. If you have a male, you will find the smaller, white *testes*. Above the place where the stomach was, is the large, saclike swim bladder. What is its function? _____

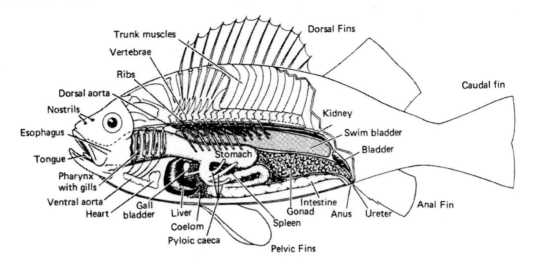

Dorsal to the swim bladder are the long, dark-red *kidneys* lying along the spine. Trace the *ureter* posteriorly to the *urinary bladder*, just above the anus.

Carefully remove the skin from the area just below the gills and open the *pericardial cavity* containing the heart, which is almost "in the fish's mouth." Locate the large *cardinal vein* carrying blood from the body into the heart. The blood enters the heart through the *sinus venosus* (sī′nəs•vĭ•nō′səs), a sac on top of the heart. From here it goes into the upper chamber, the *atrium*, then into the lower *ventricle*. The ventral aorta leaves the heart (it has a large, muscular bulb at its base), carrying blood forward to the gills.

Hold your fish in the normal swimming position, with the head pointed away from you. Cut the skin away from the skull and gradually scrape away the skull bone with your scalpel. When it gets very thin, you may pick the bone away with your forceps, revealing the *brain*. In the far front are the *olfactory lobes*. A short distance behind these are the lobes of the *cerebrum*, followed by the very large *optic lobes*, and finally by the *cerebellum*. The cerebellum extends posteriorly to the *medulla* (no distinct separation) and the *spinal cord*. Draw and label the parts of the brain that you expose.

DISCUSSION:

1. What structures that are found in fish are similar to those of animals that you have studied previously? _____

_____

2. What structures are unique? _____

_____

EXERCISE 6-8
DISSECTION OF THE FROG

INTRODUCTION:

Frogs are the most familiar of all amphibians. Most are aquatic or semi-aquatic and are frequently seen in or around every small pond. The leopard frog (*Rana pipiens*), which is most often supplied for dissections, is often found in grassy fields. Eggs are laid in masses of hundreds, with each egg surrounded by a jellylike protective coating. If you look in shallow water, you can usually find these egg clumps during the spring months. They hatch into tadpoles, which are almost all head, plus a tail. It may be very interesting to raise some tadpoles in an aquarium while they metamorphose into the adult frog.

PROBLEM:

To become familiar with the anatomy of the frog, and to note differences from the anatomy of other organisms.

MATERIALS:

Preserved frog specimen; dissection tools; wax dissection pan.

PROCEDURE:

A. **External Anatomy**

Before you begin your dissection, take a look at the external features of your frog. Note the stubby front legs and the large hind legs. Can you give a reason for the variation in size? _____

_____

Near the front of the head are the paired *nostrils*. Behind these are the *eyes*. Notice that the eyes protrude above the rest of the head. What advantage does this have for the frog? _____

_____

Just behind the eyes are the circular *tympanic membranes*. These are the external ears of the frog. In the green frog and the bullfrog, the male has a larger tympanum than the female. Also in some species, in the breeding season, the male has an enlarged thumb that helps it clasp the female while mating. Try to determine the sex of your frog.

B. **Mouth**

Pry open the mouth of your specimen. It may be necessary to break or cut some of the tissue and bones at the back corner of the jaws to do this. Examine the frog's *tongue*. Where is it attached? _____ Why is it attached in this manner? _____

_____

_____

On the lower jaw in the center of the back you will find a slitlike *glottis*. This is the opening to the lungs. If you have a male, you might be able to find the *vocal sac openings* at the corners of the lower jaw.

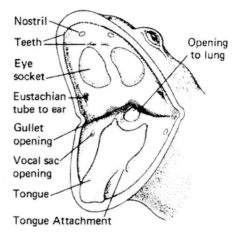

Just above the glottis, locate the openings of the *esophagus*. On the sides of the roof of the mouth are the openings to the *eustachian tubes*. These allow for an equilibrium of pressures on the tympanum. In the front of the upper jaw are the *internal nares*. These are openings from the nostrils on the outside. The frog breathes air into his closed mouth through the nostrils and then, by contracting the oral cavity, he forces air down through the glottis to the lungs.

Frogs have two sets of teeth. The *maxillary teeth* are on the edges of the upper jaw (maxilla), and the *vomerine* (vō'mə•rīn) *teeth* are on the roof of the mouth just behind the nares. These serve in helping the frog to hold its prey in its mouth before swallowing.

C. Muscular System

First, you will examine the muscular system of the frog. Lay the frog on its abdomen in your dissection pan and pin the extended appendages down. Lift the loose abdominal skin and with your scissors cut through the skin only, cutting along the mid-dorsal line from the anus to the back of the head. Now cut slits across to each of the legs so that the skin can be folded back like two swinging doors. Carefully remove the thin skin from the frog and compare its muscles to those shown in the figure at the right.

Describe the various muscles.

_____

_____

_____

_____

_____

_____

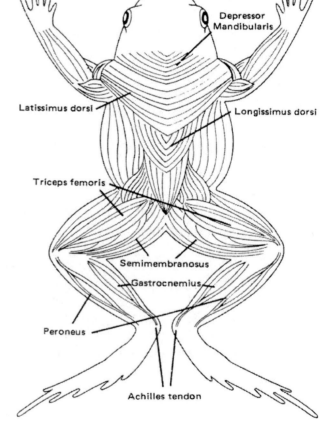

_____

_____

_____

_____

D. Internal Anatomy

Lay the frog on its back in your dissection pan and pin the extended appendages down. Lift the loose abdominal skin and, with your scissors, cut through the skin only, cutting along the mid-ventral line from the anus to the chin. Now cut slits across to each of the legs so that the skin can be folded back like two swinging doors. Notice the large number of blood vessels in this skin. What is their purpose? _____

_____

_____

Now repeat the same cutting procedure with the muscles of the abdomen. Cut slightly to one side of the center line to avoid destroying the abdominal vein that lies just below the muscles at this point. When you reach the thorax, you will have to cut through the bones of the *pectoral girdle* (shoulder). Be careful not to cut so deeply as to destroy organs underneath. Fold the muscles back and pin them out of the way.

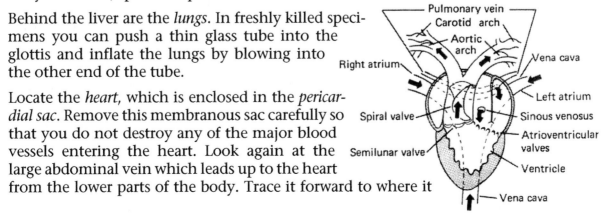

If you have a female in breeding condition, you may find almost the entire body cavity to be filled with tiny black-and-white *eggs*. Remove these so that you may examine the rest of the organs.

The most prominent of all organs is usually the lobed *liver* in the anterior part of the body cavity. If you lift this up, you will see attached to its dorsal side the greenish *gall bladder*. The liver is the largest of the digestive glands; secretions from it empty into the gall bladder for storage. On the left side and somewhat below the liver is the *stomach*. Cut this open and examine the folds on the inside, called *rugae* (rū′gē). Leading to the anterior part of the stomach is the short *esophagus*. After the food is partly digested, it leaves the stomach and enters first the *small intestine* and then the *large intestine*. From here, the feces passes into the *cloaca* and on through the *anus*. The intestines and the stomach are all attached by a thin, sheetlike membrane, the *mesentery* (mĕ′zən•tĕr•ē). On this mesentery near the curve of the stomach is the flat, yellow *pancreas*, and not far away is the red, spherical *spleen*.

Behind the liver are the *lungs*. In freshly killed specimens you can push a thin glass tube into the glottis and inflate the lungs by blowing into the other end of the tube.

Locate the *heart*, which is enclosed in the *pericardial sac*. Remove this membranous sac carefully so that you do not destroy any of the major blood vessels entering the heart. Look again at the large abdominal vein which leads up to the heart from the lower parts of the body. Trace it forward to where it becomes the *posterior vena cava* just before entering the heart. The *right* and *left anterior vena cava* join this to form the saclike *sinus venosus* which flows into the *right atrium*. Blood from here is pumped into the single *ventricle*, from where it is then pumped through the *conus arteriosus*, which you can find on the ventral surface of the heart. Here blood can flow in three directions. If it goes to the *pulmocutaneous* (pŭl′mə•cū•tā′nē•əs) *arch* it passes to the lungs and to the skin to pick up oxygen, and then by means of the *pulmonary veins* back to the heart into the *left atrium*. Here it again goes to the ventricle and out the conus arteriosus. The other two routes are the *carotid arch* to the arms and head, and the *aortic arch* to the rest of the body.

Remove the heart by cutting off these major blood vessels, and cut it in half on a longitudinal plane parallel to the dorsal and ventral surfaces. Locate each of the three chambers of the heart and examine their interiors. Why are the walls of the ventricle thicker than those of the atrium? _____

_____

Why are the atria smaller than the ventricle? _____

_____

Attached to the back of the body cavity and near the vertebral column are the two *kidneys*. These often have yellow, fingerlike *fat bodies* attached to them. Find the *ureter* which leads from each kidney down to the *urinary bladder*. If you have a female, you have already found and removed the egg-filled ovaries. Locate the *oviduct* leading from these to the cloaca. If you have a male, search for the oval testes near the kidneys.

DISCUSSION:

1. Give a possible explanation for the presence of the rugae in the frog's stomach. _____

    _____

2. The frog is often used in high school biology labs for dissections because its basic internal anatomy is similar to that of the human. Evolutionists say that this is because all vertebrates have a common ancestry. How does a creationist explain this? _____

    _____

    _____

3. We have now seen several animals in the lab that can regenerate certain parts of their bodies. Tadpoles can regenerate parts, but the adult frog is very limited in this ability, as are human beings. List as many animals as you can that are capable of some regeneration. _____

    _____

EXERCISE 6-9
THE ANATOMY OF FLIGHT

INTRODUCTION:

Flying has intrigued man for as long as he has watched birds soaring in the evening breezes. Many ancient legends are recorded about man's early attempts at flying, such as the Greek legend of Daedalus (dē'dəl•əs) and his son Icarus (ĭ'kə•rəs).[16] Today it would seem that man no longer needs to marvel at flying. He has already flown in jets faster than the speed of sound, and has traveled through space to the moon and back. Despite his great achievements, however, man has yet to match the amazing skill and grace of the birds and other flying animals.

PROBLEM:

To study the mechanisms and adaptations for flight; and to compare these mechanisms and adaptations in the birds, mammals, and insects that fly.

MATERIALS:

Empty spool; rubber tubing; index card; tape; pin; sheet of notebook paper; skeletons and specimens of birds, bats, and insects (optional); bird feather; microscopic bones from mammals and birds; lower leg from freshly killed chicken or pheasant.

PROCEDURE:

A. Flight Experiment

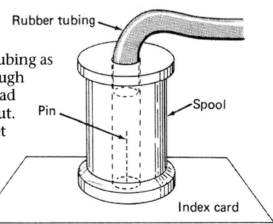

Prepare an empty spool with a piece of rubber tubing as shown in the figure at the right. Push a pin through the center of a 3x5 index card and tape the head against the card so that the pin will not fall out. Set the card on the table with the pin up and set the hole of the spool over the pin so that the pin moves freely inside it. Blow *into* the rubber tubing as hard and as long as you can, and while doing this slide the card off of the edge of the table or lift the spool.

What happens?

You might get the same results by sucking air in through the rubber hose, but it is important here that you are blowing *into* the tubing, hence *down* onto the index card.

Now take a sheet of paper and hold one edge with both hands. While holding the paper just *below* your mouth, blow across the top of it. What happens? _____

---

16. Daedalus is the legendary builder of the Cretan labyrinth; he made wings to enable himself and his son Icarus to escape imprisonment. (For more information on this legend, visit <http://www.loggia.com/myth/daedalus.html>.) For an fascinating Web site on labyrinths, visit <http://www.labyreims.com/e-index.html> and click on **Site plan** for a detailed navigation table.

The above two experiments exemplify "Bernoulli's principle," which states that as a fluid (liquid or gas) flows faster, its pressure is reduced. Bernoulli's principle is involved in airplanes, too. Their wings are shaped like an airfoil (*see figure at right*).

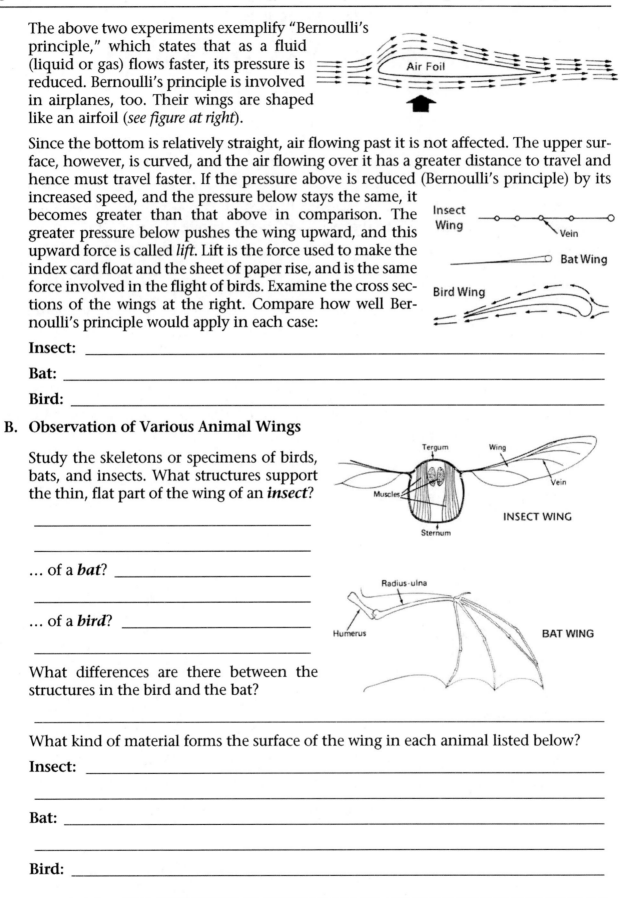

Since the bottom is relatively straight, air flowing past it is not affected. The upper surface, however, is curved, and the air flowing over it has a greater distance to travel and hence must travel faster. If the pressure above is reduced (Bernoulli's principle) by its increased speed, and the pressure below stays the same, it becomes greater than that above in comparison. The greater pressure below pushes the wing upward, and this upward force is called *lift*. Lift is the force used to make the index card float and the sheet of paper rise, and is the same force involved in the flight of birds. Examine the cross sections of the wings at the right. Compare how well Bernoulli's principle would apply in each case:

Insect: _____

Bat: _____

Bird: _____

B. **Observation of Various Animal Wings**

Study the skeletons or specimens of birds, bats, and insects. What structures support the thin, flat part of the wing of an ***insect***?

_____

_____

... of a ***bat***? _____

_____

... of a ***bird***? _____

_____

What differences are there between the structures in the bird and the bat?

_____

What kind of material forms the surface of the wing in each animal listed below?

Insect: _____

_____

Bat: _____

_____

Bird: _____

_____

What is unique about the insect wing as compared to the others? List several characteristics, if possible. _____

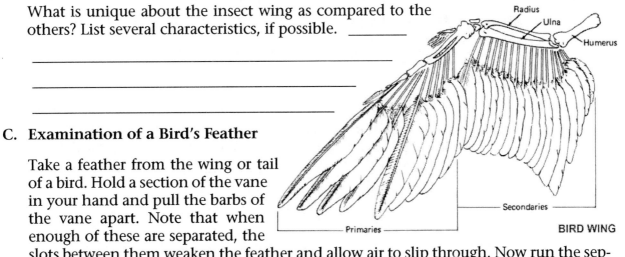

Radius
Ulna
Humerus
Secondaries
Primaries
BIRD WING

_____

_____

_____

C. Examination of a Bird's Feather

Take a feather from the wing or tail of a bird. Hold a section of the vane in your hand and pull the barbs of the vane apart. Note that when enough of these are separated, the slots between them weaken the feather and allow air to slip through. Now run the separated barbs between your fingers until they are "magically" reconnected. How does a bird reconnect separated barbs? _____

_____

_____

Examine a section of the vane under a microscope. Draw a picture of what you see, and label the *rachis* (rā'kəs), *barbs*, *barbules*, and *hooks*.

D. Flight Features of a Bird

Flying requires at least that the animal is light in weight and that it has a large surface area, such as the wings. List as many of the features as possible that are found in birds to aid them in flying. _____

_____

_____

Some other animals—such as the flying squirrel, the flying lemur, and the flying lizard of Southeast Asia—appear to have flying abilities. In what major way are their abilities different from those of birds and bats and flying insects? _____

_____

What characteristics might you find in the flying squirrel that help it to fly? _____

_____

_____

E. **Comparison of Animal Bones**

Examine two bones from a mammal and two bones from a bird. One of the bones from each of these animals should be broken in half. Describe the similarities and differences between these sets of bones. _____

_____

_____

Why do these differences exist? _____

_____

Examine the leg from a freshly killed chicken or pheasant (or other large bird). If you cut the leg off just above the ankle joint (the one that looks like a backward knee), you should be able to find a tough tendon. Holding the lower leg in one hand, pull on this tendon with the other hand. What happens? _____

How is this arrangement beneficial to the bird? _____

_____

_____

DISCUSSION:

As you have seen in this exercise, different kinds of organisms fly.

1. How do evolutionists explain this fact? _____

_____

2. Is this explanation scientific? _____ Explain your answer. _____

_____

3. How do creationists explain this fact? _____

_____

4. Is this explanation scientific? _____ Explain your answer. _____

_____

EXERCISE 7-1
THE SKELETAL SYSTEM

INTRODUCTION:

Bones are the structures of your body that provide support, shape, protection of organs, and attachment for leverage of your muscles. Everyone is familiar with bones, at least with those in the meat we eat. Many people consider bones to be dead or nonliving things, but they are completely wrong in this assumption, as you will discover in this exercise.

PROBLEM:

To determine the structure and composition of bones and to become familiar with the human skeletal system.

MATERIALS:

Prepared slide of bone cross section; acetic acid; Bunsen burner; two bones; human skeleton (or charts of it); any other available skeletons.

PROCEDURE:

A. **Observation of Bone and Bone Tissue**

Obtain a slide of bone tissue from your teacher and examine it under the microscope. Your teacher may ask you to draw what you see. Be sure to compare the slide with the figure at the right. What parts are in the cross section of a bone that tell you that it is living tissue? _____

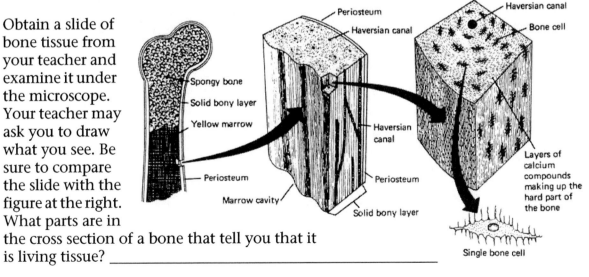

_____

Place a bone in vinegar or acetic acid so that it is fully covered. Let it remain submerged for about a week, but examine it daily. Describe the condition of the bone at the end of a week. _____

_____

Take another bone and burn it over a Bunsen burner until no further changes take place. Describe this bone. _____

_____

Use your text or another reference work to learn about the elemental composition of bone. List these elements and tell which of them were left after each of the two above experiments. _____

_____

B. Observation of the Human Skeleton

The human skeleton normally consists of 206 bones, many of which are paired. Study the chart of the human skeleton at the right or from some other source. (*You may use a 3-D model, if one is available.*) Name two sets of bones that are used in the protection of vital organs.

_____

_____

Name two ball-and-socket joints. _____

_____

_____

Give the location of one hinge joint. ___

_____

_____

Where in the body do you have a pivot joint? _____

_____

Compare the human skeleton with any others that are available. Note the differences and try to give reasons for them in terms of function. _____

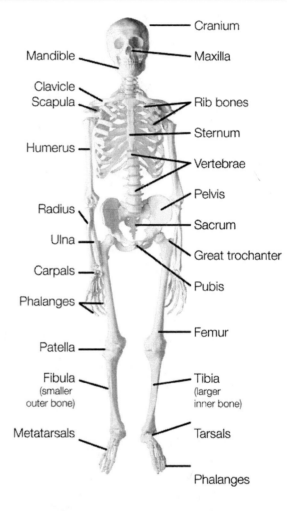

_____

_____

_____

_____

_____

_____

EXERCISE 7-2
MUSCLES

INTRODUCTION:

Muscles are used regularly in our daily lives. Every movement of our body, even blinking an eye, requires contraction of muscles. The beating of our heart involves a muscle. People are concerned about being fatigued and worn out, or about getting muscles into condition so that they do not fatigue as easily. We eat muscle in the form of steak or hamburger or other types of meat. In this exercise we will study muscle contraction in the frog, and look at the structure and types of muscles in the human body.

MATERIALS:

Double-pithed frog; dissection tools; blotter paper; frog ringer solution; bromthymol blue; hypodermic syringe; dry cell (6 volt); copper wire; prepared slides of skeletal muscle; cardiac muscle and smooth muscle tissue; microscope; charts of the human muscular system.

PROCEDURE:

A. Dissection and Testing of Frog Muscles

Have your teacher provide you with a double-pithed frog. A frog's nervous system is undeveloped enough so that when both the spinal column and the brain are destroyed (double-pithed), the frog's tissues still remain "living" for some time. Remove the skin from the calf of the hind legs and cut out the large calf muscle (*Gastrocnemius*). Be sure to include with it the tendon at its origin on the thigh and the tendon at its insertion on the foot. If possible, keep the *sciatic nerve* attached to it. This nerve continues from the dorsal side of the thigh on up to the spinal column. Place the muscle on a blotter paper wetted with frog ringer solution to prevent it from drying out. Inject the muscle with bromthymol blue solution.

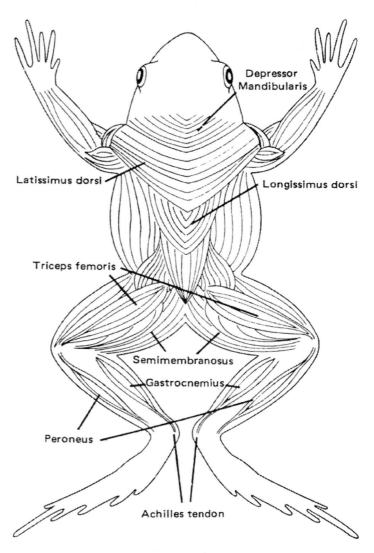

Depressor Mandibularis

Latissimus dorsi

Longissimus dorsi

Triceps femoris

Semimembranosus

Gastrocnemius

Peroneus

Achilles tendon

Now attach two copper wires to the poles of a dry cell and touch the other two ends to the nerve (if it is still attached) or to the muscle itself. What happens when the wires touch the muscle? _____

_____

Continue to touch and retouch the wire to the nerve or muscle at short intervals.

What eventually happens to the bromthymol blue? _____

_____

B. Observation of Human Muscle Tissue

Examine prepared slides of the three types of muscle tissue. Make a drawing of each type. What differences do you see between skeletal muscles, smooth muscles, and cardiac muscles? _____

_____

_____

_____

_____

C. The Human Muscular System

Examine the drawings of the human muscular system. Almost every muscle has another muscle that works against it. The latter is called an *antagonistic muscle.*

What is the antagonist for the *biceps*? _____

_____

_____

What is the antagonist for the *rectus femoris*? ____

_____

_____

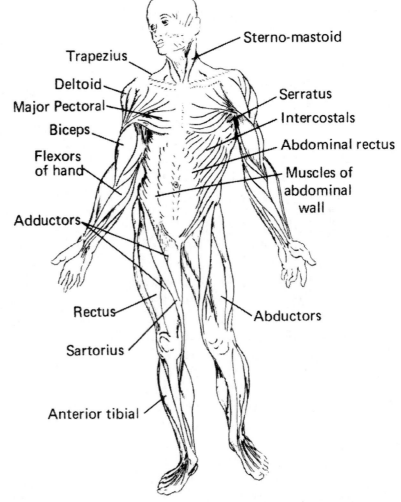

Trapezius

Deltoid

Major Pectoral

Biceps

Flexors of hand

Adductors

Rectus

Sartorius

Anterior tibial

Sterno-mastoid

Serratus

Intercostals

Abdominal rectus

Muscles of abdominal wall

Abductors

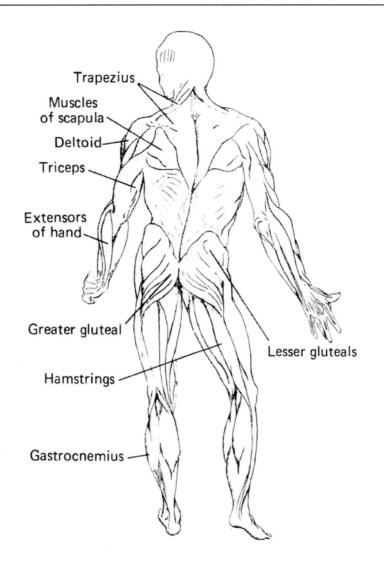

Trapezius

Muscles
of scapula

Deltoid

Triceps

Extensors
of hand

Greater gluteal

Lesser gluteals

Hamstrings

Gastrocnemius

DISCUSSION:

1. Review the reaction of the bromthymol blue when the muscle was repeatedly stimu-
   lated. _____

2. What is bromthymol blue a test for? _____

3. Write a short paragraph on the chemistry of muscle reactions, and include in it an
   explanation for the reaction of the bromthymol blue. _____

   _____

   _____

   _____

   _____

   _____

EXERCISE 7-3
CIRCULATION OF BLOOD

INTRODUCTION:

You have probably been told that blood travels away from the heart in *arteries*, passes through smaller vessels called *arterioles*, into tiny *capillaries*, then into the *venules*, and finally through the *veins* back into the heart. Only in relatively modern times has man known these basic facts. The purpose of this exercise is to demonstrate this circulation for you.

PROBLEM:

To view the methods of circulation of blood in vertebrates.

MATERIALS:

Prepared slides of artery and vein cross sections; microscope; 2 large beakers; hot and cold water; live guppy or frog; paper towels; petri dish; cork board for frog; pins.

Note how the fingers should be pressed on the vein.

PROCEDURE:

A. Measuring Heart Rate

The radial artery, used for taking your pulse, is between the base of the thumb and the two large tendons in the middle of your wrist. Use your first two fingers to detect your pulse. If you use your thumb, you will have difficulty feeling your pulse correctly, as it too has its own pulse that will interfere with a correct reading of the one in your wrist. The pulse is a surge of blood flowing through an artery. Why does it come in rhythmic surges rather than in one continuous flow? _____

_____

Count your pulse rate for one minute, while sitting, and record it. _____

An adult male has an average resting heart rate of about 72 beats per minute, and a woman has between 76 and 80. It is higher for teens (84), children (90), newborns (130–140), and babies in the womb (140–150). In elderly people, it is lower (50–65).

What is your pulsebeat after heavy exercise for one minute? _____

Compare this with the increase in other family members or friends and try to account for the differences in the increase. _____

_____

After passing through the capillaries, blood pressure is greatly reduced. In fact, often the contraction of your skeletal muscles surrounding the veins is needed to help squeeze the blood back to the heart. People who are required to stand motionless in one place for a long time often wiggle their toes and flex their leg muscles. If this is not done, they may faint, as not enough blood gets back to the heart and, therefore, also not to the brain. What keeps the blood from being squeezed backward into the arteries, rather than forward to the heart? _____

If you have veins on your arms, legs, feet, or hands that bulge out prominently, you can see how this is done. With two fingers, push down on the vein, and then draw one finger along the vein towards the heart for several inches. Lift the finger nearest to the heart (keeping the other one in place all this time). Does the blood flow backwards to the first finger? _____ This is due to one-way "butterfly" valves that allow the blood to flow toward the heart, but not in the reverse direction.

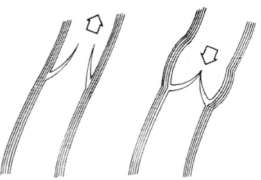

Vein with Valve Open and Closed

Arteries are thick-walled, elastic, and muscular, so that they can stretch with the surging pulses of the blood. Veins are thinner and nonmuscular. Examine a cross section of an artery and of a vein under the microscope. Draw a picture of each of them, making sure to identify them.

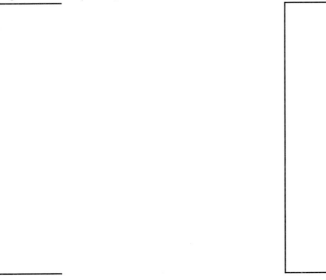

Obtain two large beakers or other large containers. Fill one with ice-cold water and the other with hot water (not so hot as to burn). Place one hand into each container and leave them there for one minute. Remove your hands and observe their color. Record your observations.

Explain why this difference exists. _____

_____

B. Flow of Blood in Capillaries

You can observe the actual flow of blood in the capillaries of a guppy's tail or in the web of a frog's foot.

If you use a fish, select a small one and wrap it in wet absorbent cotton or paper toweling, covering the anterior portion of the body, leaving only the tail exposed. Cover the bottom of a shallow petri dish with water, then place the wrapped fish in the water. Lay a glass slide over the tail to hold it out flat and examine it under the microscope.

With a frog, wrap all but one hind leg in wet cheesecloth or strong paper toweling. Be sure that it is kept moist constantly, and that the frog is immobile except for the one hind leg. Use a cork board with a circle cut out near one end, and tape or tie the frog down to it. Tie the free leg down firmly so that the webbing is over the hole. Once the leg is secured, you can pin the webbing in place, or tie thread around the toes and stretch them out. (*This will not harm the frog.*) The leg must be fully immobilized, otherwise the frog may tear his webbing. Now you can observe the capillaries of the webbing through the microscope.

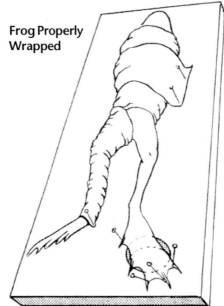

Frog Properly Wrapped

In the capillaries, the red corpuscles will be flowing in single file. If you find a place where they are not, it is either an arteriole or a venule. Do any of these larger vessels pulsate? _____ Is the direction of flow toward larger or smaller vessels? _____

_____

Try to locate some white blood corpuscles. They are much less numerous than the red corpuscles. Describe how the white corpuscles differ from the red. _____

_____

DISCUSSION:

1. How could you tell the difference between arterioles and venules? _____

   _____

   _____

2. Describe the difference between your drawing of arteries and that of veins. _____

   _____

EXERCISE 7-4
HUMAN BLOOD AND BLOOD TYPING

INTRODUCTION:

Blood has long been recognized for its importance. Early man may not have known what blood does, but he did know that the loss of it was related to death. The Egyptians at one time even used blood from animals as a fertilizer for their crops, suspecting that it had some magic life-substance in it. Today we know that blood is not magic, but it is still very important.

Blood typing is discussed in chapters 7 and 16 of the text. Many scientists believe that if every blood protein were analyzed, we would find that each person has a unique blood type. In this exercise we will examine some of the properties of human blood, and each student will analyze his blood to determine its type, using just one category, including types *A*, *B*, *AB*, and *O*.

PROBLEM:

To familiarize yourself with blood smears and to discover the importance and techniques of blood typing.

MATERIALS:

Hemoglobin paper; hemoglobin chart; sterile lancet; toothpicks; alcohol; cotton ball; blood-testing serums Anti-*A* and Anti-*B*; Wright's stain[17]; Ringer's solution[18]; distilled water; petri dish; 3 slides; microscope.

PROCEDURE:

> *What is Forward Typing?*
>
> In forward typing a sample of the blood is mixed with serum that contains antibodies against type *A* blood ("anti-*A* serum"). Another sample of blood is then mixed with serum that contains antibodies against type *B* blood ("anti-*B* serum"). Patterns of clotting are then observed and recorded as follows:
>
> Type *A* blood clots when mixed with anti-*A* serum.
>
> Type *B* blood clots when mixed with anti-*B* serum.
>
> Type *AB* blood clots when mixed with both anti-*A* and anti-*B* serums.
>
> Type *O* blood does not clot when mixed with either anti-*A* or anti-*B* serum.
>
> 1. Persons with type *A* blood can receive blood transfusions from donors with type *A* or type *O* blood.
> 2. Persons with type *B* blood can receive transfusions from donors with type *B* or type *O* blood.
> 3. Persons with type *AB* blood can receive transfusions from donors with type *AB*, type *A*, type *B*, or type *O* blood.
>
> **Note:** Serum that contains antibodies against type *Rh* Positive Blood ("anti-*Rh* serum") is not discussed here.

Note: *BE SURE TO READ ALL DIRECTIONS BELOW BEFORE BEGINNING.* Once you start, you will have to work quickly to avoid the drying of your blood.

Take one of your clean slides and have your teacher put one drop of each of the blood-testing serums on it. Keep each drop separate (one on each end of the slide) or your results will not be correct. The serums are usually color-coded to distinguish them. Ask your teacher to identify the color for each serum.

Now swing your left hand downward to force blood into it (your right hand if you are left-handed). Swab the index finger of your left hand with alcohol to clean the skin. Remove the sterile lancet from its wrapper, being careful not to contaminate it, and

---

17. Wright's stain is named after American pathologist, James H. Wright (1871–1928). Two of the important components of Wright's stain are the bluish dye *hematoxylin* ($C_{16}H_{14}O_6$) and the pinkish dye *eosin* ($C_{20}H_8Br_4O_5$). Hematoxylin has a basic pH and stains nuclei and some cytoplasmic granules dark blue. Eosin has an acidic pH and stains some granules and other materials pink.

18. Ringer's solution is named after British physiologist Sydney Ringer (1835–1910). It consists of fresh boiled distilled water containing 8.6 gm sodium chloride, 0.3 gm potassium chloride, and 0.33 gram calcium chloride per liter—the same concentrations as found in body fluids.

make a quick jab of your index finger with the lancet point. *Under no condition should the lancet be used by more than one person.* You will discover that pricking your finger by this method is easy and relatively painless. Remember, you have approximately twelve pints of blood in your body, and you need only five drops of it for this exercise.

Once you have pricked your finger, wipe the first drop away, as it usually contains mostly serum. Place one drop of blood next to each of the colored serums on your slide. Place a third drop near one end of the second slide to use for microscopic examination (see below), and a fourth drop on the hemoglobin paper. Once you have collected enough blood, place a cotton swab soaked in alcohol on your finger to aid in stopping the bleeding, if necessary, and to insure freedom from bacteria.

Next, quickly follow this procedure to prepare your second slide. Take another slide and draw it toward the blood drop at an angle of thirty degrees until it touches it and the drop spreads out along the edge of the slide by capillary action. Immediately pull the blood along behind the upper slide as you move it rapidly across the lower slide. This will give you a smooth, thin layer of blood, which you can now let dry.

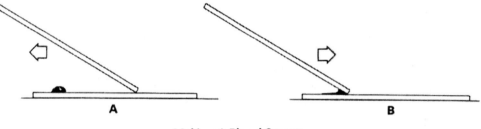

Making A Blood Smear

Return to your first slide. Being careful to avoid mixture of the two different types of testing serums, mix each serum with its adjacent drop of blood, using separate toothpicks. Now observe for a moment to see if clumping of cells is taking place. Record here which blood and serum mixture clumped (if any) and continue on with the rest of the experiment before computing your blood type.

Was there a reaction with Anti-*A* serum? _____

... with Anti-*B* serum? _____

Now take your second slide, with the dried blood smear on it, and place it over a petri dish, covering the blood with Wright's stain. Let it stand for 1 to 3 minutes, then gradually dilute with distilled water until it is diluted to about half. Let it stand again for 2 to 3 minutes. Wash all this off with distilled water until it appears lavender-pink. Let it dry thoroughly, and then inspect it under the microscope. (If an oil-immersion lens is available, you can examine your blood under this without any cover slip. Merely place the oil directly on the stained blood. You should be able to find not only the red corpuscles, but also the various types of white blood cells. These will have granules that are deep blue or bright red, while others are lilac-colored.)

Blood Type	Plasma Factor Present	Red Corpuscle Factor Present
A	Agglutinogen *B*	Antigen *A*
B	Agglutinogen *A*	Antigen *B*
AB	no factors	Antigen *A* and *B*
O	Agglutinogens *A* and *B*	no factors

While you are waiting for your blood smear to dry, or between slide-staining procedures, you can work on determining your blood type. The table above shows the contents of each type of blood. Notice that there can be both *agglutinogens* and *antigens* in the blood. The agglutinogens are within the plasma of the blood, and the antigens are protein factors which may or may not be in the red corpuscles themselves. Agglutinogen *A* will agglutinate (clump) red blood corpuscles containing antigen *A*, but will not affect antigen *B*. The reverse is true for agglutinogen *B*. Therefore, in any transfusion, any combination of bloods with matching agglutinogens and antigens must be avoided. Examine the table above carefully and compare your blood test with those of others in your class. Then complete the following Agglutination (*clumping*) Table.

Blood Type	Anti-*A* Serum Agglutinates	Anti-*B* Serum Agglutinates
A		
B		
AB		
O		

After you have determined your blood type, check it by cross-matching your blood with that of another person. Repeat the procedures for drawing blood given above. Place a drop of your blood on one end of a slide, and a drop from someone else with the same blood type on the other end of the slide. Dilute both of these with a small drop of human ringer solution and then mix the two bloods. Do you detect any reaction? _____

What other mixtures might be made to test the accuracy of your first test? _____

_____

Take the hemoglobin paper that you put the drop of blood on earlier. Compare it with the hemoglobin "color" chart at the right. This will tell you the approximate percentage of hemoglobin in your blood. What is the hemoglobin percentage for your blood according to the chart?

If it appears too low, the low reading could be due to some error in your technique, or it could mean that you are anemic, and you should check with your doctor.

Hemoglobin Chart

RED: "Stop and Think": > 10.0% (R)

YELLOW: "Be Careful": > 7.0% (Y)

GREEN: "Good Going": < 7.0% (G)

HbA1c		Average Blood Sugar
16.0	RED	420
15.0	RED	390
14.0	RED	360
13.0	RED	330
12.0	RED	300
11.0	RED	270
10.0	RED	240
9.0	RED	210
8.0	YELLOW	180
7.0	YELLOW	150
6.0	GREEN	120
5.0	GREEN	90

DISCUSSION:

1. What is your blood type? _____

2. What agglutinogens and antigens are present in your blood? _____

    _____

3. From what types of blood could you safely receive a transfusion? _____

    _____

    Why? _____

    _____

4. Find out how many people in your family have each blood type and record percentages here.

    Type *A* _____ Type *B* _____ Type *AB* _____ Type *O* _____

    Compare these ratios with the following table.

    ### Percentage of Blood Groups in Various Populations

People Group [a]	Type O	Type A	Type B	Type AB
Euro-American	45%	40%	11%	4%
Afro-American	49%	27%	20%	4%
Native American	79%	16%	4%	1%
Asian-American	40%	28%	27%	5%

    a. Information in this chart is based on data from <http://www.bloodbook.com/world-abo.html>.

EXERCISE 7-5
RESPIRATION

INTRODUCTION:

Breathing has always been known to be an important process. Its importance is even reflected by many words in our vocabulary. We talk about being *inspired* (breathe in), or *aspiring* to a position (breathe towards). When a man dies, we say that he *expires* (breathes out). You have already examined the respiratory systems of several animals that you have dissected in class. Recall the spongy appearance and texture of the lungs. Refer to the drawing in your textbook concerning the structure of the human respiratory system (Figure 16-11, page 204).

PROBLEM:

To examine the functioning of the human respiratory system.

MATERIALS:

240-ml beaker; bromthymol blue; hydrogen peroxide; glass tubing; Erlenmeyer flask (or conical flask); rubber cork; Bunsen burner; prepared slides of the respiratory tract; microscope; paper bag; balloons; spirometer; KOH or NaOH crystals.

PROCEDURE:

A. Breath Experiments

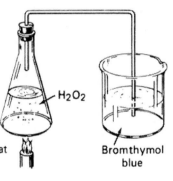

Take a 250-ml beaker and put 25 ml of bromthymol blue into it. Insert a glass tube and breathe into the solution through it. What happens? _____

Now take an Erlenmeyer flask and put 50 ml of hydrogen peroxide into it. Cover it with a cork that has a glass tube coming out of it, as shown in the figure at the right. Heat the flask gently, while the other end of the tube is in a beaker with 25 ml of bromthymol blue solution. What happens now? _____

The equation for the heating of hydrogen peroxide is:

$$2H_2O_2 \xrightarrow{\text{Heat}} 2H_2O + O_2$$

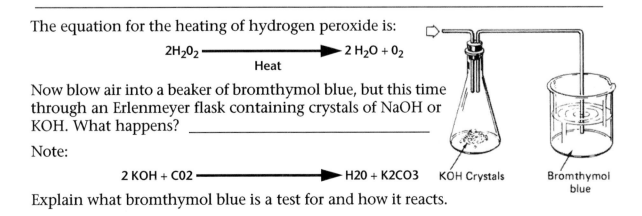

Now blow air into a beaker of bromthymol blue, but this time through an Erlenmeyer flask containing crystals of NaOH or KOH. What happens? _____

Note:

$$2KOH + CO_2 \longrightarrow H_2O + K_2CO_3$$

Explain what bromthymol blue is a test for and how it reacts.

_____

B. Microscope Work

Examine the slides of the respiratory tract under your microscope. Look for the hairlike cilia on the linings of the nasal passages and trachea. What is their function? _____

_____

_____

C. Breathing Test

Have someone sit quietly for a moment; count and record the number of times that he breathes in one minute. _____

Now have him *hyperventilate* (breathe in and out deeply) for one minute. He should be careful when doing this. Be sure that he is sitting down, and he should stop if he begins to feel faint. After hyperventilating, have him stop and return to normal breathing. Start immediately to count his respiration rate and record it here. _____

Now have him cover his mouth and nose with a paper bag and breathe into and out of this for two minutes. Count his respiration during his second minute of breathing into the bag, and for the first minute after he stops breathing into the bag. Record the results here. _____

D. Spirometer Test

Partially fill a four- or five-quart can with water, allowing it to float low, upside-down in a tub of water. Blow through a rubber tube whose open end is under the can. This will force air into the inverted can, causing it to rise. You can make a computation of your lung volume by marking the level of the can both before and after blowing into it. Later remove the can and turn it right-side-up. Then fill it with water to the first mark. Now gradually fill it to the second mark, measuring the volume that you pour in with a graduated cylinder. Record your lung capacity here: _____

Spirometer

DISCUSSION:

1. How do you explain your results from recording the respiration rate while breathing into a bag? _____

   _____

2. What determines the rate of one's breathing? _____

   _____

3. Devise an experiment that will enable you to prove your explanation for questions 1 and 2. Describe your experiment here. If possible, try it out and record your results. Discuss these in class with your teacher. _____

   _____

   _____

   _____

   _____

   _____

   _____

4. Compare your lung capacity with that of the other members of your family. Can you find any relationship between the lung capacity and a person's size, or his physical fitness, or any other trait? _____

   _____

5. For what reasons are there inaccuracies in the results shown by your inverted-can spirometer? _____

   _____

   _____

EXERCISE 7-6
ENZYME ACTION IN DIGESTION

INTRODUCTION:

In this exercise we want to examine the function of digestive enzymes. The human digestive system has over ten different enzymes; however, we will study only one in this exercise, since all the enzymes work in the same basic manner.

PROBLEM:

To determine the functioning of digestive enzymes.

MATERIALS:

Bunsen burner; ring stand; one large pyrex and one small pyrex beaker; graduated cylinder; thermometer; 6 pyrex test tubes; a small bead of plastic; 30 ml of Lugol's iodine solution; 30 ml of Benedict's solution; 50 ml of 1% starch solution.

PROCEDURE:

A. **Saliva and Starch Test 1**

Place a clean piece of plastic in your mouth to induce salivation. Collect about 10 ml of saliva in a small beaker.

Fill two test tubes with 5 ml of 1% starch solution. Place these in a water bath (*see figure at the right*), which is kept at body temperature (98°F, or 37°C). After about 10 minutes, remove them from the heat and add 5 ml of Lugol's solution to one, and 5 ml of Benedict's solution to the other. Heat them, one at a time, over the burner. Be sure to use a pyrex test tube pointed in a safe direction while moving it back and forth over the flame. Record your results in the table below.

Next fill two test tubes with 5 ml of starch solution and add 5 drops of saliva to each. Test these every two minutes for ten minutes. For a final comparison, test saliva alone in a test tube with 5 ml of Benedict's solution. Record all of your results below.

Test Tubes and Contents	Results of Test
1. Starch tested with Lugol's	
2. Starch tested with Benedict's	
3. Starch and Saliva with Lugol's	
4. Starch and Saliva with Benedict's	
5. Saliva with Benedict's	

Give an explanation of your results. _____

_____

_____

B. Saliva and Starch Test 2

Now prepare 6 pyrex test tubes with 5 ml of the starch solution and add 5 drops of saliva to each of them. Place two tubes in an ice bath, keep two tubes at room temperature, and put two tubes into the water bath at 110°F (43.3°C). Incubate all of these for 20 minutes. Test 2 tubes of each temperature, one for starch with Lugol's solution and the other for sugar with Benedict's solution. Record your results in the table below.

Effect of Temperature on Salivary Digestion of Starch

Condition	Temperature	Test	Results
Ice water		Lugol's	
Ice water		Benedict's	
Room temperature		Lugol's	
Room temperature		Benedict's	
Water bath	110°F	Lugol's	
Water bath	110°F	Benedict's	

Explain your results. _____

_____

_____

_____

_____

_____

_____

EXERCISE 7-7
SPECIAL SENSES

INTRODUCTION:

Because of our senses, we are aware of many wonderful things. People who have lost their sight or hearing have learned to appreciate this fact, but most of the rest of us fail to appreciate our senses fully. In this lab we will examine some of the senses that we use to explore our environment.

PROBLEM:

To determine in some special ways our ability to smell, taste, touch, and see.

MATERIALS:

Potato; onion; apple; vinegar solution; salt solution; aspirin solution; sugar solution; water; toothpicks; blindfold; 2 probes; ice water bath; hot water bath; clean sheet of paper; water-soluble ink.

PROCEDURE:

A. Taste Tests

Blindfold a friend and have him hold his nose. Take small cubes of raw potato, onion, and apple and, with a toothpick, place them one at a time in his mouth. Have him chew them separately and test how well he can determine what he is eating. Why is he not more successful? _____

Now prepare five solutions:

1. tap water (control)
2. 1 part vinegar to 2 parts water
3. 10% NaCl solution
4. aspirin solution
5. 5% sugar solution.

Apply each of these separately to your friend's tongue with a toothpick, but do not let him know which solution you are using. Apply these to different areas of the tongue to locate where he can detect each taste. Be sure that he rinses out his mouth after each test. Record your results by shading in the sensitive areas for each solution on figure below.

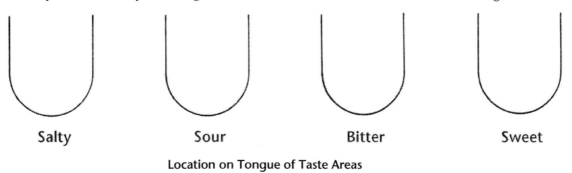

| Salty | Sour | Bitter | Sweet |

Location on Tongue of Taste Areas

B. Touch Tests

Again blindfold your friend. Using two probes, attempt to discover his ability to tell whether you are touching the back of his hand with one or two probes. You will find out that within a certain distance two probes feel like only one, and only farther apart do they feel like two. Record this distance for the back of his hand. _____

Record the similar distance on his finger tip. _____

Cool one probe in an ice bath. Move this around on the back of the hand of your blind-folded friend. Notice that only certain areas of his hand are sensitive to the cold. In an area of one square inch on the back of his hand, locate all spots sensitive to the cold probe and mark them with a small spot of water-soluble ink.

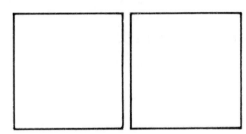

Now take a probe that has been heated in a water bath until it is hot, without becoming so hot as to burn. Do the same as you did with the cold probe. In the one-inch squares at left mark the spots for cold receptors and for heat receptors, respectively.

Are heat and cold receptors in identical locations on your hand? _____

What do you conclude about temperature receptors? _____

_____

C. Sight Tests

Near the center of a clean, unlined sheet of paper, draw two dark spots about the size of a penny and about three inches apart. Look at the two dark circles from as close as you can without feeling uncomfortable and without the spots becoming blurry.

Now close your right eye and stare at the right dot with your left eye. Gradually move farther back from the paper as you continue to stare at the right dot. You will reach a point where the left dot seems to disappear. Record this distance from the paper to your eye. _____

Try this again, staring at the left dot with your right eye. From your knowledge of the structure of the eye, explain why this blindspot exists. _____

_____

Why don't you normally have trouble with this blindspot? _____

_____

DISCUSSION:

1. What effect might the blindspot have on a driver who has vision in one eye only?

_____

_____

2. Should he be allowed to drive? _____

_____

EXERCISE 8-1
STRUCTURE OF PLANT STEMS

INTRODUCTION:

Stems function to support and display leaves to sunlight, and to transport nutrients and raw materials to both the leaves and the roots. Trees and shrubs have strong, woody stems, whereas herbaceous stems, like those of peas, can hold up little more than their own weight. What specializations can we find in stems that result in such variety of form? _____

_____

_____

PROBLEM:

To determine the similarities and distinguishing characteristics of a woody dicot stem, an herbaceous dicot stem, and a monocot stem.

MATERIALS:

Prepared slides of basswood cross section; buttercup stem cross section; corn stem cross section; sample of woody tissue; dilute nitric acid; potassium chlorate; iodine solution; sulfuric acid (65%); phloroglucinol; 75% HCl; test tubes; petri dishes.

PROCEDURE:

A. Preparation of Woody Elements or Cells

Cut a bit of woody material into small pieces. Place in a test tube and cover with nitric acid (HNO_3). Add a few crystals of potassium chlorate ($KClO_3$) and heat gently. Allow material to react from 4 to 5 minutes. Pour material into a petri dish and wash with tap water. Remove pieces of wood to a slide and separate woody fibers with a dissecting needle. Add a drop of phloroglucinol and allow to dry. Add a drop of 75% HC1 and cover with a cover slip. Observe on both low and high power. Draw the kinds of cells that you see. The large, hollow, cylindrical cells are called vessels, and the narrow, thread-like cells are called *fibers*. *Tracheids* are long, tapering cells with walls thickened by spirals or rings of cellulose. Label appropriately the cells you have drawn.

B. The Woody Dicot Stem

Study a woody stem cross section on low power. In the very center locate the *pith*. These cells are very large and thin-walled. What is their function? _____

_____

Locate the rays radiating outward. How far do they extend? _____

_____

What is their function? _____

_____

Outward from the pith, identify the series of circles called *annual rings*. These mark the growth of each season. Use high power to examine the wood. On the basis of cell size and wall thickness, what kinds of cells would you say make up the *xylem*? _____

_____

At the outer edge of the xylem locate the *vascular cambium*. What two tissues does it produce? _____

Outside the cambium, *phloem tissues* alternate with pith rays that are part of the *cortex*, which lies outside the phloem tissues. What is the function of the cortex? _____

_____

The phloem consists of a variety of cells. *Sieve tubes* are large, thin-walled cells containing protoplasm but no nuclei. *Companion cells* are smaller, thin-walled cells that do have nuclei. The companion cells are thought to control transport in the sieve tubes. Masses of cells making very thick cell walls are *phloem fibers*. Outside the cortex is another layer of cambium called the *cork cambium*. What is the function of the cork?

_____

Why must it be continually produced? _____

_____

Below are cross sections of a woody stem and an herbaceous stem with the areas of various tissues drawn. Differentiate between different kinds of cells on the basis of size, shape, and wall thickness. Label the following: *pith, pith ray, cortex, cork, cork cambium, xylem, phloem, vascular cambium,* and *annual ring*.

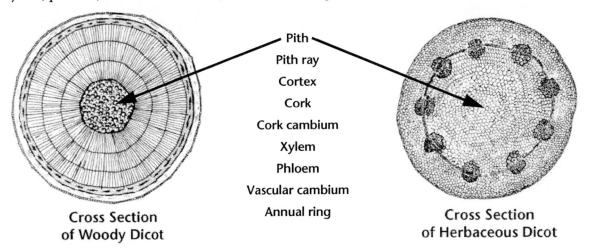

Pith
Pith ray
Cortex
Cork
Cork cambium
Xylem
Phloem
Vascular cambium
Annual ring

Cross Section
of Woody Dicot

Cross Section
of Herbaceous Dicot

C. **The Herbaceous Dicot and Monocot Stems**

Examine a cross section of buttercup stem on low power. Move the slide so you view all portions of the stem. The outer surface is covered with epidermis. How many cell layers does it have? _____

Inside the epidermis are large, loosely packed cortex cells. Within the cortex are fibrovascular bundles. What kinds of tissue make up these bundles? (Hint: consider cell size and shape.) _____

How are the many fibrovascular bundles arranged in your stem cross section? _____

_____

In the cross section of the herbaceous dicot, identify the tissue of its stem and show the cellular detail in one fibrovascular bundle.

Examine all regions of a monocot stem cross section on low power. What do you notice about the cell walls of the outer layers of cells? _____

_____

This layer is called the *rind*. How does the arrangement of the fibrovascular bundles in the monocot stem differ from the arrangement of them in the dicot stem? _____

_____

Switch to high power and focus on a single fibrovascular bundle. What kind of tissue seems to be lacking? _____

Notice that thick-walled cells completely surround the fibrovascular bundle. These are *sclerenchyma* (sklə•rĕng′kə•mə) *cells*. They provide mechanical support, but no conduction. Draw below a monocot fibrovascular bundle on high power and label: *xylem, phloem, sieve tubes, companion cells, sclerenchyma cells,* and *intercellular space*.

DISCUSSION:

1. What kinds of tissues are present in the stems of all plants? _____

    _____

    _____

2. Some plant stems have chlorophyl in the cells of the cortex. What other plant function may be carried on by the stem? _____

    _____

    _____

EXERCISE 8-2
ROOT STRUCTURE

INTRODUCTION:

Plant roots function to anchor a plant and absorb nutrients. Roots may also store food and provide for transport of some materials.

PROBLEM:

To determine the arrangement of root tissues and relate this to their functions.

MATERIALS:

Tap root (carrots); germinating radish seeds; prepared slides of buttercup root cross sections; prepared slides of longitudinal section of onion root tip; microscope.

PROCEDURE:

A. **Root Tip Regions and Primary Tissues of Young Root**

Examine a longitudinal section of a root tip through low power. The very tip of the root consists of an irregular mass of cells called the root cap. What is the function of the root cap? _____

Immediately behind the root cap is a *meristematic* (mĕr′əs•tə•mă′tĭk) *region*, or region of cell division. This is the source of the "primary" or first tissues in the root. Notice that these cells are small and cubical in shape. Farther up the tip is the region of elongation. How much longer are these cells than those in the meristematic region? _____

As cells in the region of elongation lengthen, what happens to the root cap and meristematic regions? _____

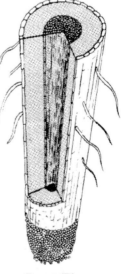

Beyond the region of elongation notice that the cells differentiate. This is the region of *maturation*, where the cells are specialized for doing particular jobs. At the right is an enlarged drawing of a root tip. Label: *root cap, meristematic region, elongation region, maturation region, root hair, cortex, epidermis,* and *vascular tissue.*

B. **The Root Hairs**

Examine the roots of a very young radish seedling. What is an important function of the root hairs? _____

What is the advantage of having a large number of root hairs?

_____

_____

_____

Root Tip

C. The Primary Tissues of a Mature Fibrous Root

Study a cross section of a buttercup root and identify the kinds of tissues it has. A single outer layer of cells is called the epidermis. Inside the epidermis is the cortex. How many cells thick is the cortex? _____

The center of the root contains the vascular tissues (*xylem* and *phloem*). Xylem consists of large, hollow, thick-walled cells. Phloem cells are smaller and thin-walled. These are separated from the cortex by the *pericycle* (pĕr'ə•sī•kəl), which produces secondary roots; and the *endodermis*, which controls the passage of materials from the cortex into the central cylinder (xylem and phloem). The various tissues are shown in the diagram below. Label: *epidermis, cortex, central cylinder, phloem, xylem, cambium, endodermis,* and *pericycle.*

D. Secondary Tissues of Mature Tap Root

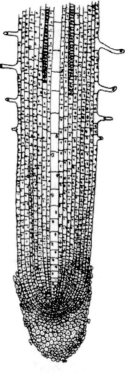

Mature Fiberous Root Primary Tissues

Study the external features of a small carrot. The diameter of a root is the result of cell division in the vascular cambium. The epidermis and cortex have disintegrated. Notice the very small stem to which the leaves are attached. Locate several *secondary roots* that grow laterally. With a scalpel cut the carrot longitudinally into two halves. Make a sketch of the carrot and identify and label these parts: *periderm* (a tough outer layer produced by the pericycle), *secondary phloem* (food conducting tissue), *secondary xylem* (principal water-conducting tissue), and *secondary roots.*

DISCUSSION:

1. Briefly describe the function of the following tissues:

   epidermis _____

   cortex _____

   xylem _____

   phloem _____

   meristematic tissue _____

   vascular cambium _____

2. Why is it a poor practice to transplant vascular plants after pulling them out of the ground? _____

   _____

EXERCISE 8-3
LEAF STRUCTURES

INTRODUCTION:

One of the most important of all biological processes—*photosynthesis*—occurs in leaves. In this activity we will learn how a leaf is ideally suited for this task.

PROBLEM:

To determine the structure of the leaf.

MATERIALS:

Prepared slides of leaf epidermis and leaf cross sections; samples of various leaf types (geranium, coleus, and lilac); slides and cover slips; microscopes; 10% salt solution.

PROCEDURE:

A. The External Form of the Leaf

Examine a whole leaf, including the leaf stalk called the petiole. Describe the shape of the leaf. _____

What proportion of the leaf is penetrated by veins? _____ What are the major tissues present in the petiole and the veins? _____

Draw a sketch of your leaf.

Label: *petiole, midrib, veins, margin,* and *blade.*

B. The Leaf Epidermis

Remove a small bit of epidermis from the lower side of a fresh lilac leaf by tearing through the blade and twisting slightly. You should obtain a transparent layer or sheet of cells. Place this transparent layer in a drop of water on a slide. Examine on low power. The openings in the epidermis are called *stomates.*
The bean-shaped cells that surround the stoma are *guard cells.* What structures can you see in the guard cells? _____

Place a drop of 10% salt solution at an edge of the cover slip and draw it through the mount by blotting up water from the other side. Describe what happened to the guard cells. _____

_____

Now add distilled water to the slide and draw it through. What are the results? _____

_____

_____

Sketch here the section of the epidermis on low power and label: *epidermal cell, stomates, guard cell,* and *chloroplast.*

C. The Internal Structure of a Leaf

Study a prepared slide of the cross section of a leaf. Compare the upper and lower epidermis. How are they different from each other? _____

_____

The cells of the epidermis are covered by a waxy layer called the cuticle. Between the upper and lower epidermis we find the *mesophyll* (mĕ'zə•fĭl). The upper part of the mesophyll consists of tightly packed rows of cells, called the *palisade layer.* The lower mesophyll is called the *spongy layer* because it contains many air spaces. Within the cells of this layer we also find the veins of the leaf. Compare the palisade and spongy cells.

_____

_____

_____

Which cells have more chloroplasts? _____

Which cells are more directly exposed to light? _____

What seems to be the function of the spongy cells? _____

_____

Locate a section of the leaf that shows a vein, the mesophyll, and stomates clearly. Draw this section of the leaf and label: upper and lower epidermis, mesophyll, palisade layer, spongy layer, spongy cells, air spaces, stomates, a chloroplast, vein, guard cell, and cuticle.

DISCUSSION:

1. Describe the pathway and the process by which CO_2 in the atmosphere reaches the palisade layer. _____

   _____

   _____

2. Describe how water reaches the palisade layer of the cells. _____

   _____

   _____

   _____

   _____

3. Water lily leaves grow on the surface of the water. Where would you expect to find stomates on this leaf? _____

   Why? _____

EXERCISE 8-4
PHOTOSYNTHESIS

INTRODUCTION:

If you were to ask someone, "What is the most important biological process in the world?", what would he answer? Growth, movement, or breathing might appear in a list of possible answers; but let us consider photosynthesis. All the food that we eat, the oxygen we breathe, many building materials, most of our clothing, and many medicines are supplied by green plants. However, even with all the benefits that green plants supply, 10% of the world's people (800 million) go to bed hungry every night. Part of the solution to the problem of world hunger may lie in our understanding of this basic food-making process—*photosynthesis*. In this lab activity we will consider factors that affect photosynthesis and try to determine the conditions that are necessary for photosynthesis to take place.

PROBLEMS:

1. To determine some of the characteristics of chlorophyll and its role in photosynthesis.
2. To establish the necessity of having CO_2 available for photosynthesis.
3. To determine the kind and amount of light necessary for photosynthesis.
4. To determine the effect of temperature on photosynthetic rates.

A. Chlorophyll and Photosynthesis

MATERIALS:

Coleus plants; black opaque paper; methyl alcohol; fine sand; acetone; petroleum ether; test tubes; corks; paper clips; filter; paper toweling; Lugol's iodine; geranium plants; glass rods; hot plate; cellophane; beakers; asbestos pad; petri dish.

PROCEDURE:

Remove a leaf from a coleus plant that shows green and white markings. Make a drawing of this leaf and show the green and white areas by different shading. Boil the leaf in alcohol on a hot plate until it is colorless.

Note: Alcohol will burst into flames very easily. *Have no open flames in the room while working.* If it does catch fire, cover the beaker with an asbestos pad to smother the flame.

The chlorophyll has now been removed. Dry the leaf on paper toweling and place it in a petri dish and flood with Lugol's iodine. After a few minutes remove and dry the leaf. What color now appears on the leaf? _____

125

On the previous page, make a drawing of the stained leaf next to the drawing of the leaf containing chlorophyll. Show, by shading, the areas with and without starch. (Starch stains blue in the presence of iodine.) Explain the pattern that you see. _____

_____

Cover a green portion of a leaf of a coleus plant (front and back) with strips of black paper. Use paper clips to hold the paper in place. On another leaf of the same plant cover a green portion with strips of clear cellophane. Place this plant in bright sunlight. It will be used later.

In the above steps it has become evident that chlorophyll is important in producing food (starch). Now we will remove the pigments from some leaves and learn more about them by *chromatography*. This is a technique used to separate a mixture of substances. The substances to be separated are placed at the same spot on filter paper, and a solvent is added. The solvent moves through the filter paper by capillary action. As the solvent moves through the spot, the substances dissolve in the solvent and are carried along with the solvent at different rates. The resulting "graph" is a *chromatogram*.

Place geranium leaves in boiling water for 2 minutes. Transfer the leaves to a beaker of 100 ml of alcohol boiling on a hot plate. Allow to boil for 5 minutes. Why did the alcohol become green, and why did the water not become green? _____

_____

_____

_____

_____

While this chlorophyll solution is cooling, set up materials to make chromatograms of your chlorophyll solution. We will use two techniques of chromatography—*column chromatography* and *disk chromatography*.

Column Chromatography

Insert a paper clip into the bottom of a cork. (*See figure at right*.) Attach a strip of filter paper so the paper will be $\frac{1}{2}$ inch from the bottom of a test tube. Place a pencil mark $1\frac{1}{2}$ inches from the tip of the filter paper. Using a glass rod, spread a drop of the chlorophyll solution on the pencil mark. Wait for the drop to dry before adding another one. Keep adding the chlorophyll solution until you have a heavy, almost black line. Now pour 5 ml of acetone and 5 ml of petroleum ether into the test tube. Put the filter paper into the test tube so that it reaches into the solvent and the cork fits tightly. Be sure that the filter paper does not touch the sides of the test tube. Watch the solvent move up the filter paper and notice the separation of the pigments. Remove your chromatogram as

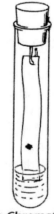

Column Chromatography
Set-Up

soon as the solvent reaches the top of the filter paper. Mark the position and color of the various pigments. Attach your column chromatogram in the space provided. List the order of pigment spots you observed: _____

_____

_____

Disc Chromatography

Cut a wick in a circular piece of filter paper. (*See figure at right.*) Place a heavy spot of pigment at the fold. Pour 5 ml of acetone and 5 ml of petroleum ether into a petri dish. Place the filter paper over the petri dish with the wick in the solvent. Observe the separation until the solvent reaches the edge of the filter paper and then remove. Is the order of the pigment separation the same as you obtained before? _____

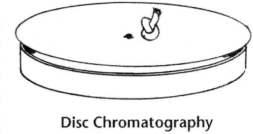

**Disc Chromatography
Set-Up**

_____

How many different pigments are there in the solution? _____

B. Carbon Dioxide and Photosynthesis

Carbon dioxide is present in the atmosphere at a very low concentration of 0.035%. We might question the importance of such a small amount of CO_2 in the process of photosynthesis.

MATERIALS:

Geranium plant; 5.0% baking soda solution; phenolphthalein; 0.04% sodium hydroxide; 2 wide-mouth quart jars with tight-fitting lids.

PROCEDURE:

Remove two healthy leaves from a geranium plant and place each leaf into a 100-ml beaker of water. Into 1 large glass container with a lid, add a 5.0% solution of baking soda to a depth of 1 inch. Set this container aside and label it "A." Into another large glass container place about 1 inch of tap water and label this container "B." Add 3 to 5 drops of *phenolphthalein solution* to each jar. (Phenolphthalein is an indicator that is pink in basic solutions and colorless in acid solutions. When CO_2 is dissolved in water, it produces carbonic acid.) To solution "B" add one drop at a time of 0.04% NaOH solution. Stop when the solution becomes pink. This indicates that Na has combined with CO_2, removing it from the water.

Why was the baking soda solution added to container "A"? _____

_____

Why was the NaOH solution added to container "B"? _____

_____

Place one of the beakers containing the geranium leaf into each glass container. Tightly close the lids and set in bright light for 2 days (or place under artificial light for 24 hours). Test both of these leaves for the presence of starch. Which leaf has a measurable amount of starch? _____

C. Light and Photosynthesis

As you have already learned, light energy is necessary for the process of photosynthesis. Let's learn how much light and what kind of light are best for photosynthesis to take place.

MATERIALS:

Coleus plants prepared in section A; Lugol's solution; elodea (ĭ•lō′dē•ə, a small, submerged, aquatic herb); sodium bicarbonate; wood splints; test tubes; funnels; beakers; glass rods.

PROCEDURE:

Remove the coleus leaves that you covered with black paper and cellophane. Make a sketch of each of these, showing the distribution of color and the position of the black paper and cellophane. Boil the leaves in alcohol on a hot plate. Remove and dry the leaves. Place them in a petri dish and flood with Lugol's iodine solution. Remove after 2 minutes and make a sketch of each leaf and show the distribution of starch. Are there any green areas that did not have starch in them? _____ Explain. _____

_____

Force sprigs of elodea into a 3-inch funnel. Place the funnel upside down in a beaker of water that has had 2 gm of sodium bicarbonate added. Fill a test tube with water and place your thumb over the top. Invert the test tube and lower it into the beaker so that the test tube fits over the funnel stem and the opening is under water. Set up another beaker exactly the same way. Place one beaker in bright light and the other beaker in darkness for 2 days and then observe.

What change has taken place? _____

_____

Test any gas produced in these test tubes for oxygen. This is done by placing a glowing (not burning) splint into the gas. If the gas is oxygen, the splint will burst into flame.

Now that we have established that oxygen is produced in photosynthesis, we may also determine the rate of photosynthesis under varying light conditions.

Cut 4 different 3-inch sprigs of elodea under water. Tie the tip of each sprig to a 2-inch piece of glass rod. Fill 4 test tubes to within one inch of the top with water and add a pinch of sodium bicarbonate. Put a sprig of elodea with attached glass rod into each test tube. (The glass rod will keep the elodea at the bottom of the tube.) Place these test tubes in 4 different light conditions for 15 minutes. Then count the number of bubbles produced per minute by each plant. Five minutes later make another count of the number of bubbles produced per minute. Record your results in the following chart.

Light Condition	Trial 1 Bubbles/minute	Trial 2 Bubbles/minute	Average Bubbles/minute

What effect does light intensity have on photosynthesis? _____

_____

D. Temperature and Photosynthesis

If for some reason the earth were farther away from the sun, what effect would this have on the rate of photosynthesis? (Or, if you were above the Arctic Circle and had months of continuous sunlight, at what temperature ranges would you find photosynthesis to be fastest?) This activity will give some clues.

MATERIALS:

Elodea; 100-ml graduated cylinder; 1000-ml battery jar; high-intensity lamp or projector; ice; hot water; centigrade thermometers; sodium bicarbonate.

PROCEDURE:

Obtain a 3-inch sprig of elodea and attach the tip to a glass rod. Fill a 100-ml graduated cylinder almost full of water and add a pinch of sodium bicarbonate. Place the elodea sprig into the 100-ml graduated cylinder. Set this cylinder inside a 1000-ml battery jar and fill the battery jar 3/4 full of water at room temperature. Suspend a thermometer into the battery jar to keep a constant check on the temperature. Position a strong light source close to the beaker. Add ice to the battery jar to bring the temperature down to 10°C. When bubbles appear, record the number per minute and the temperature. Add hot water to the battery jar to raise the temperature 5°C. Shake all previous bubbles off the sprig and again record the number of bubbles per minute and the temperature. Repeat until the temperature reaches 40°C. Prepare a graph of your results. Place the temperature on the horizontal axis, and the bubbles per minute on the vertical axis.

EXERCISE 8-5
WATER MOVEMENT IN PLANTS

INTRODUCTION:

One of the most important raw materials in photosynthesis is *water*. In the photophase of photosynthesis, water molecules are split and electrons and hydrogen are separated and used in the photosynthetic process. Water is also important in the transport of raw materials in plants. But how can water reach heights of 300 feet in giant trees? In this lab activity we will demonstrate some of the forces at work in moving water in plants.

PROBLEM:

To illustrate the forces that bring about the movement of water in plants.

MATERIALS:

Lima bean seeds; wax paper; glass rod; beakers; 100-ml graduated cylinders; capillary glass tubing; fresh celery; colored water; potted plants; clear plastic bags; clear nail polish or collodion solution; cobalt chloride paper; petri dish; large carrot; cork borer; 1-hole rubber stopper; paraffin; molasses; ring stand; battery jar; paper clips; hot plate; ruler.

PROCEDURE:

A. **Imbibition**

Dry materials such as seeds are able to absorb water into their tissues, greatly increasing their volume. Place 100 ml of water into a beaker and add 25 lima bean seeds. Cover with a petri dish. Let stand for 24 hours. Pour off the water into a graduated cylinder. How much water remained after 24 hours? _____

Explain. _____

_____

B. **Capillarity: Cohesion and Adhesion**

Place a small drop of water on a piece of waxed paper. Touch the drop with a glass rod and slowly move the tip of the glass rod, keeping it in contact with the paper and the water. Describe the shape and movement of the water. _____

_____

Heat a piece of glass tubing until the glass is softened. Quickly pull the ends of the glass rod apart. The heated portion is drawn into a very thin glass tube called a capillary tube. Remove the 4- to 5-inch center portion of the capillary tube and place one end of the tube into a petri dish of colored water. Explain what you observe in terms of adhesion (attraction between molecules of 2 different substances) and cohesion (attraction between like molecules). _____

_____

_____

C. Water Movement in Stems and Leaves

Cut the basal 2 inches of a stalk of fresh celery off while holding the lower end under water in a 1000-ml beaker or battery jar. Cutting under water prevents air bubbles from forming in the xylem vessels. Add red food coloring to give the water a distinct color. Observe the celery in 24 hours. Make a cross-sectional sketch to show the distribution of the food coloring.

D. Transpiration

It has been said that 95% of the water that enters a plant's roots is lost by the plant's leaves. This lost water is seldom seen. The present demonstration is used to obtain a visible amount of water from a plant.

Water a healthy potted plant, such as a geranium. Place a plastic bag over the leafy portion. Tie the opening closed around the stem. Place the plant in bright light for 24 hours. Record your observations. _____

_____

E. Diffusion

Diffusion is the movement of molecules from an area of high concentration to an area of low concentration. When this takes place through a living membrane, it is called *osmosis*. This can be demonstrated with the carrot.

With a cork borer, remove a $\frac{1}{2}$ to $\frac{3}{4}$-inch diameter portion from the center of the carrot. Remove as deep a core as possible without splitting the carrot. Fill the carrot almost full of molasses. Insert a glass tube 18 inches long into a one-hole rubber stopper. Push the rubber stopper carefully into the carrot. Seal the junction of the carrot and rubber stopper by coating it with several layers of melted paraffin. Place the carrot into a battery jar. Cover the carrot with water and support it with a clamp on a ring stand.

Record the height of molasses in the glass tube. _____

What is the height after 24 hours? _____

After 48 hours? _____

Why does the level of the molasses rise in the tube? _____

_____

F. Transpiration and Stomates

Do a study of the varying numbers of stomates on the lower surface of a leaf and the proportionate rate of water loss. Select 5 different kinds of plants. Coat the undersides of a leaf of each plant with a layer of clear nail polish or collodium. Let dry and remove the layers of polish. This

Glass tube

Stopper sealed with paraffin

Carrot

Molasses

Water

Diffusion Experiment

layer now has a print of the cells of the lower epidermis. Label these prints for later microscopic observation. Attach a blue cobalt chloride strip and a pink cobalt chloride strip to the lower surface of a leaf on each of the 5 plants. Blue cobalt chloride paper changes to pink as it picks up moisture. Record the time required for the blue paper to become the same color as the standard pink one.

Back in the laboratory, observe the epidermis prints and record the number of stomates in the field of vision on high power. Record all your observations on the chart below.

	Plant Name	Number of Stomates/Field	Time Required for Color Change
1.			
2.			
3.			
4.			
5.			

DISCUSSION:

What relationship exists between the number of stomates and the rates of transpiration? _____

_____

EXERCISE 8-6
SEED GERMINATION

INTRODUCTION:

Most of the plants from which we gain food and raw materials are grown from seeds. Thus, it is very important to know what conditions or treatment a seed needs if it is to germinate and grow into a healthy plant.

PROBLEM:

To study germination in monocots and dicots.

MATERIALS:

Seeds of corn, beans, oats, clover, radishes, squash, cucumbers, castor beans, pine, sorghum, and tomatoes; wide-mouth quart jars or beakers; sawdust; paper towels.

PROCEDURE:

A. Comparing Germination of Monocot and Dicot Seeds

Line a beaker with several layers of paper toweling. Fill the inside of this with sawdust. Place a row of alternating corn and bean seeds between the paper toweling and the glass beaker. Moisten the sawdust and place the beaker in a warm, dark place. Observe daily for seven successive school days and make a sketch of the development of each kind of seed. Label: *cotyledons, radical,* and *seed coat.*

B. Conditions Necessary for Germination

Prepare 4 germinators as in Part A. Number the germinators 1, 2, 3, and 4. Select 3 kinds of seeds from those available and obtain 100 of each kind. Place 25 seeds of each kind that you selected into 4 germinators. You will then have 75 seeds (25 each of three kinds) in each germinator. Add water to germinators 1, 3, and 4 (keep No. 2 dry). Place your germinators in the following conditions, and observe daily for a week, and fill in the information in the data table at the top of the next page.

135

Germination Conditions	Seeds Used	Total Number of Seeds Germinated Each Day							Total Number of Seeds Used by the Study Group	Percentage of Total Germinated Seeds
		1	2	3	4	5	6	7		
1. light, moist, warm										
2. dark, dry, warm										
3. dark, moist, cool										
4. dark, moist, warm										

DISCUSSION:

1. What conditions proved best for plant germination? _____

    _____

    _____

2. Did some seeds germinate (in any of the conditions) better than others? _____

    Explain. _____

    _____

    _____

EXERCISE 8-7
FLOWERS AND FLOWER STRUCTURE

INTRODUCTION:

The angiosperms or flowering plants carry on sexual reproduction. This requires the production and union of sperm and egg. How does the flower carry out this function?

PROBLEM:

To determine the structure and function of the flower.

MATERIALS:

Preserved lily flowers; fresh or preserved flowers of other types; fresh pollen; scalpels; forceps; sucrose solution; slides; cover slips; microscopes; Vaseline™; pollen from early plant collections; crystalline violet or methylene (mĕ′thə•lēn) blue stain.

PROCEDURE:

A. Structure of a Typical Monocot Flower

The lily is an example of a *complete* flower. Its parts appear in 3s or groups of 3; hence, it is a monocot. Study your preserved lily. The tip of the stem to which the flower is attached is called the *receptacle*. The lily appears to have 6 *petals*, but the outer ring of 3 are really *sepals* (which are green in most flowers and leaflike). Collectively the sepals are called the *calyx* (kā′lĭks), and the inner ring of true petals is called the *corolla*. These floral parts are not absolutely necessary for reproduction to take place; so they are called *accessory parts*.

The long, slender *stamens* are inside the ring of petals. They consist of a slender *filament* and the expanded *anther* which produce the pollen. How many stamens are there? _____ In the very center of the flower is the *pistil*. It is divided into the *ovary, style,* and *stigma*. The ovary is the expanded base of the pistil; from its top extends the slender style, and at the top is the slightly expanded and sticky stigma. The stamen and pistil are the *essential parts* of the flower, since they are necessary for reproduction.

Make a sketch of your flower and label: *receptacle, sepals, petals, stamen, pistil, filament, anther, stigma, style,* and *ovary*. Inside the ovary there are tiny, immature seeds called *ovules*. Find them by cutting the ovary in half longitudinally with a scalpel.

B. Other Floral Types

The lily is an example of a *perfect* (having all four floral parts) and *complete* (having both essential parts) flower.

Sketch and label two other flowers and indicate whether these flowers are complete or incomplete and whether they are perfect or imperfect.

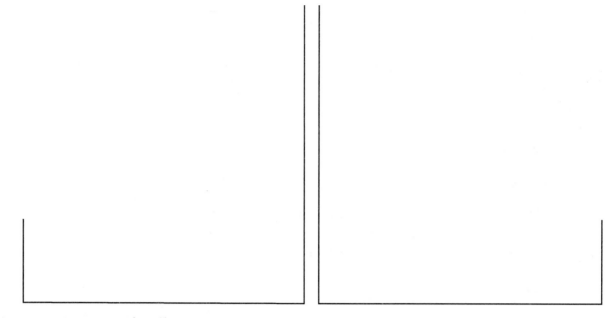

C. Examination of Pollen Types

Transfer a sample of pollen to a drop of water on a slide and add a drop of crystal violet or methylene blue stain. Allow to stain for 3 to 5 minutes. Examine the grains under low power and high power, if necessary. The part of the pollen grain that is visible is the outer protective covering called the *exine* (ĕk′sēn). The exine of each kind of pollen is specific in its shape, much as each person's fingerprints can be used for identification purposes. The exine may have distinct ridges, furrows, spines, or wings. Draw the exines of five different kinds of pollen grains, showing the distinctive features of each kind and give the name of each kind of pollen.

_____ _____ _____ _____ _____

D. Germinating Pollen Grains

If ripe pollen grains are placed in a sugar solution, they can be germinated in the class room. Make a hanging drop mount of ripe pollen, using 10% sugar solution. Some pollen tubes may develop within 20 minutes, and others may require 1 to 2 days. Observe different stages of pollen growth and see if you can identify the nuclei present in the pollen tube. Explain your findings. _____

_____

_____

DISCUSSION:

1. What are the "male" parts of the flower? _____

   _____

2. What are the "female" parts of the flower? _____

   _____

3. List as many ways as possible that are used to transfer the pollen from the anther to the stigma. _____

   _____

   _____

4. Some flowers can pollinate themselves, while many must be pollinated by pollen from another flower. What possible advantage might there be in cross-pollination? _____

   _____

5. Sections C and D showed how pollen from each different plant is distinct. Can you think of any way that this might be used in scientific investigations? _____

   _____

   _____

   _____

EXERCISE 8-8
MINERAL REQUIREMENTS OF PLANTS

INTRODUCTION:

All of us are dependent upon the growth of green plants. We have discovered in previous experiments that plants need water, carbon dioxide, and sunlight to produce the food (sugar) and oxygen that all animal life needs to exist. Some of you probably work in your own back-yard garden or live on a farm and, as a result, realize that plants need more than just carbon dioxide and water to survive and grow healthy. We often say that plants need good, rich soil for full growth. Actually, under special conditions, plants can be grown without soil, but they do need the minerals found in the soil. Growing plants in gravel or in solutions of water, to which the proper minerals have been added, is called *hydroponics* (hī'drə•pä'nĭks) and is sometimes the method used to produce commercially grown roses, tomatoes, and a few other plants.

PROBLEM:

To examine the mineral requirements of plants by using hydroponics.

MATERIALS:

Distilled water (6 liters); minerals described in section B; six 500-ml pyrex beakers; wax paper; bean or pea seeds; petri dishes; blotter paper; rubber bands; marking pencil; pH paper; ferric tartrate (a salt or ester of tartaric acid containing iron); 6 large, closeable containers.

PROCEDURE:

A. Germinating the Plants

Place approximately 20 bean or pea seeds between moist blotter paper inside of a petri dish, and keep them in a dark place. Germinate these until the roots are at least 4 inches long. The length of time required for germination to this stage will depend on the seed type and the temperature and humidity of the germinator.

B. Preparation of Solutions

In this experiment you will be working with six different stock solutions of varying mineral content. Mix these solutions as instructed below, being careful to make all weights as accurate as possible (use an analytical balance if available). After mixing the solutions, put them into a large container and close the container tightly so that no moisture is lost in evaporation.

Solution 1.	1 liter distilled water
Solution 2.	1 liter distilled water
	0.95 g Calcium nitrate $Ca(NO_3)_2$ $4H_2O$
	0.61 g Potassium nitrate KNO_3
	0.49 g Magnesium sulfate $MgSO_4$ $7H_2O$
	0.20 g Potassium dihydrogen phosphate KH_2PO_4

Solution 3. (Nitrogen deficient)	1 liter distilled water
	0.95 g Calcium chloride ($CaCl_2$)
	0.49 g Magnesium sulfate
	0.20 g Potassium dihydrogen phosphate
Solution 4. (Phosphorous deficient)	1 liter distilled water
	0.95 g Calcium nitrate
	0.61 g Potassium sulfate
	0.61 g Potassium nitrate
	0.20 g Potassium chloride KCl
Solution 5. (Potassium deficient)	1 liter distilled water
	0.95 g Calcium nitrate
	0.61 g Sodium nitrate $NaNO_3$
	0.49 g Magnesium sulfate
	0.20 g Calcium phosphate $Ca_3(PO_4)_2$
Solution 6. (alkaline)	1 liter distilled water
	0.95 g Calcium nitrate
	0.61 g Potassium nitrate
	0.49 g Magnesium sulfate
	0.20 g Potassium monohydrogen phosphate K_2HPO_4

Also mix a solution of 4.0 grams of ferric tartrate in one liter of distilled water.

C. Experimental Stage

Take your 6 pyrex 500-ml beakers and fill each of them about $\frac{3}{4}$ full with one of the 6 solutions. Label these with the wax marking pencil according to their solution number (1–6). Stretch a piece of wax paper across the mouth of the beaker and tie it down with string or a rubber band. Poke three small holes through the wax paper with a pencil. When the seeds that you have germinated in section A have roots 4 inches long, take them from the germinator and carefully push the roots through the holes in the wax paper. Put 2 plants in each jar, leaving the third hole to add more solution as it is needed. The roots should reach at least 2 inches down into the solution. Check this each day and add more solution as necessary.

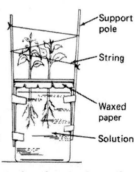

Support pole

String

Waxed paper

Solution

Hydrophonic Growth Apparatus

Every other day add 2 to 3 milliliters of the ferric tartrate solution. Place the plants where they can get sufficient sunlight and observe for four weeks. As the plants grow larger, they may need more support than just the wax paper. Tie two sticks to opposite sides of the beaker and stretch strings across between them to support the plant's stems if necessary. Keep a record of all your observations on the chart below. Test each solution with pH paper and record this also on the chart.

| Solution | Ph | Observation of Changes in Roots, Stems, and Leaves | | | |
		1st Week	2nd Week	3rd Week	4th Week
1.					
2.					
3.					
4.					
5.					
6.					

DISCUSSION:

1. Why was ferric tartrate added to each container? _____

_____

_____

2. From your experiment, list those minerals that seem to be important to plant growth, and indicate what happens to the plant that is deprived of them. _____

_____

_____

_____

3. Why was distilled water used for solution 1? _____

_____

_____

Why was it used in the other solutions? _____

_____

_____

4. From this experiment, what two functions of roots can you deduce? _____

_____

_____

_____

_____

EXERCISE 8-9
BEHAVIOR OF PLANTS

INTRODUCTION:

We do not usually think of plants as having any "behavior." Behavior is something that is generally associated only with animals and human beings; however, your text defines behavior as "the external response of the organism to environment." Plants grow, trees lose their leaves and form new ones. In addition to such basic behavior, plants do some other very interesting things.

PROBLEM:

To determine the behavior of plants in certain situations.

MATERIALS:

4 corn seeds; petri dish; blotter paper; newspaper; tape; glass marker; wire screen; 6 potted plants; ring stands, rings, and clamps; 3 cardboard boxes.

PROCEDURE:

A. Geotropism

Petri dish with 4 kernels of corn (all pointing inward)

Take four corn seeds and place them on the bottom of a petri dish, with the narrow ends of the seeds pointing toward the center of the dish. (*See figure at the right.*) Carefully cover them with two or three thicknesses of wet blotter paper. Pack enough newspaper on top of the blotter paper so that, when the lid is placed on and taped to the dish, the seeds will be pressed immovable against the bottom. Make a mark on one side of the petri dish. Place the dish on its edge, with the mark down, in a dark drawer or closet. Tape it to the side of the drawer or closet so that it will remain on edge. Check the seeds every day until they have roots and stems one to two inches long. Remove the petri dish from the dark area and examine the growth of the seeds. Do the roots and stems originate from the same part of all four seeds, or from the part that is on the top or bottom? _____

Draw the position of the four seeds and their roots and stems. Explain why they grew this way.

Select three of the potted plants. Cut three squares of wire screen large enough to cover one pot. Cut a slit in the middle of these just big enough to allow you to slip them over the plants. Be sure each pot is filled to the top with soil. Push the screen down around the edges of the pot and tie down securely with a string.

Support each pot in a ring stand—one on its side, one upside down, and one in normal position. Observe these daily for about a week. Be sure to water them a little, but always return them to their exact positions. What happens to each plant? _____

_____

_____

Explain. _____

_____

_____

_____

B. Phototropisms

Water the three remaining potted plants. Place a plant in each of three cardboard boxes. Seal the boxes shut so that they will be completely dark inside. Prepare the boxes further as follows: Do nothing more with box number 1; cut a hole the size of a quarter about one inch from the top of one side of box number 2; cut a hole the size of a quarter at about the same level as the top of the flower pot in box number 3.

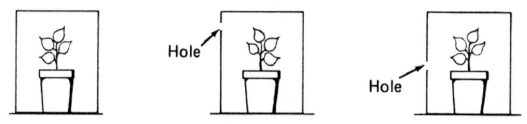

Put the boxes on a shelf with the holes facing outward toward the light. Leave them alone for at least one week; then open the boxes and observe the plants.

Describe the leaves of the plants in each box.

1. _____

2. _____

3. _____

Describe the growth of the stems in each box.

1. _____

2. _____

3. _____

Explain the results. _____

_____

_____

_____

EXERCISE 9-1
SPONTANEOUS MUTATION

INTRODUCTION:

One of the theories propounded to account for change posits spontaneous mutations among organisms. In this exercise you will use bacteria which are normally red. Bacteria are used because they will allow you to observe many generations (2–3 per hour) in a short period of time. If the bacteria mutate, you should be able to see them readily and to determine a rate of mutation in this population.

PROBLEM:

To observe spontaneous mutation in bacteria.

MATERIALS:

Stock slants of the bacterium *Serratia marcescens*; test tubes with 10 ml sterile nutrient broth; petri plates of nutrient agar (100mm); glass dally (5-inch piece of glass rod bent like a hockey stick); alcohol for flaming dally; sterile pipets (0.1 ml graduations); inoculating loop.

PROCEDURE:

A. You will inoculate a tube of nutrient broth with a small amount of bacteria from the stock slants. Incubate at room temperature for 24 hours.

   Gently suspend the cells by tapping the tube with your fingers. Aseptically place a 0.1 ml drop of the broth on the surface of the nutrient agar in a petri dish. Spread this inoculum (ĭ•nō′kyə•ləm) evenly over the surface using the alcohol flamed and cooled glass dally. Incubate for 24 hours.

   Count the total number of colonies on your plates, noting the number which are white. Also note colonies of other colors and shapes which may be contaminates.

   Explain your findings. _____

   _____

   _____

   _____

   _____

B. (Optional) Six serial 10-fold dilutions of the 24-hour-old broth culture can be made by using 9 ml sterile water blanks. This may be necessary if the colonies are too thick to count. Explain your findings. _____

   _____

   _____

   _____

   _____

DISCUSSION:

1. What was the percentage of cells that "lost" their color? _____

2. Is this number the same as the "rate" of mutation? _____

   _____

3. How might a cell lose its color? _____

   _____

4. How would you determine if this is a true mutation? _____

   _____

5. Design an experiment to "prove" your answer to No. 4. _____

   _____

   _____

   _____

   _____

   _____

   _____

6. Determine the number of cells per ml in the original broth tube. Remember: 0.1 ml
   was taken from the 10 ml broth tube. _____

   _____

EXERCISE 9-2
NATURAL SELECTION

INTRODUCTION:

The concept of survival of the fittest is illustrated when some environmental pressure acting on a population kills or inhibits the growth of most but not all of the population. In this exercise you will apply the pressure of an antibiotic on a bacterial population, using the gradient plate technique. The resulting population will have inheritable characteristics different from the original population.

PROBLEM:

Can the concept of natural selection be observed (proved) in the laboratory?

MATERIALS:

Stock slant culture of *Bacillus subtilis*; sterile petri plate (100 x 15 mm); tube of sterile nutrient agar (7-8ml/tube); tube of sterile nutrient broth (10 ml); antibiotic such as Terramycin®; inoculating loop; sterile pipette (0.1 ml graduations).

PROCEDURE:

1. Inoculate a tube of nutrient broth with a small amount of bacteria from the stock slants. Incubate at room temperature for 24 hours.

2. Pour a tube of melted nutrient agar into a petri plate and raise one edge about 1/8 inch. (Be sure agar covers the entire bottom. Of course, it will be of uneven depth.) When the bottom layer is set, place the plate on a flat surface. Onto the first layer pour a tube of melted agar containing the antibiotic (0.1 mg/ml).

Note: The concentration of the antibiotic on the surface is proportional to the thickness, which is now uneven—higher where thick, lower where thin.

Gently suspend the cells by tapping the tube with your fingers. Aseptically place a 0.1 ml drop of the broth on the surface of the antibiotic gradient. Spread this inoculum evenly over the surface using a glass pipette (that has been dipped in 95% ethyl alcohol, flamed to remove alcohol, and cooled before use). Incubate for 24 hours.

3. Select a colony that is located over the deepest part of the antibiotic layer. Repeat steps 1 and 2.

4. Increase the concentration of the antibiotic 10-fold and repeat steps 1-3.

5. After this has been accomplished, you should have a culture of bacteria that differs significantly in its tolerance of antibiotic as compared to the original culture.

DISCUSSION:

1. How do you account for the note that the antibiotic concentration is proportional to the thickness of the layer? _____

    _____

2. Does this exercise illustrate natural selection? _____

    If so, how? If not, why not? _____

    _____

    _____

3. How would you prove that the new culture is different from the original with respect to antibiotic tolerance? _____

    _____

4. How can you show that the "acquired" characteristic (resistance to the antibiotic) is inheritable? _____

    _____

EXERCISE 9-3
THE DILEMMA OF ORIGINS

INTRODUCTION:

Since the time of Charles Darwin, a great many intelligent, scientific minds have worked on refining the "theories" of evolution. Other great minds have continued to reject these "theories," and have accepted the "theory" of creation. How can intelligent men so vehemently oppose each other in scientific theory? Part of the answer is in the fact that neither "theory" can be carefully tested in the laboratory. Both are supported by only "circumstantial evidence" and by "faith." Although intellectual minds may be impressed with much of the evolutionary evidence, there are still many gaps in the "theory." This exercise is planned to examine some of these gaps and to guide the student to question them, both from the viewpoint of the evolutionist and from the viewpoint of the creationist.

MATERIALS:

None necessary other than reference works.

PROBLEMS AND DISCUSSION:

1. According to evolutionary "theories," a bat's wing evolved through a series of minor mutations that gradually enabled the bat to fly. Natural selection destroys harmful mutations and allows beneficial ones (quite rare in nature) to be maintained. In what way could "minor" mutations of the bat's ancestors' forearm be beneficial to the bat? Could these changes be of any value, until they had gone far enough to enable the bat to fly? _____

   _____

   _____

   _____

2. Evolutionists place reptiles as the ancestors of the birds. Microscopic study even reveals that reptilian scales and bird feathers begin to develop in the same manner from the same tissues. It is theorized, therefore, that feathers evolved from scales. As yet, however, no one has found an intermediate formation between scales and feathers in the fossil record. Creationists say this is because there is no link, whereas evolutionists say such a link merely has not been found. What explanation would you give for this phenomenon? _____

   _____

   _____

   _____

3. For years there have been reports concerning the presence of a large, furry, human-like creature in the wilderness areas of northern California, Oregon, Washington, and British Columbia. Sightings of the creature itself, or its footprints, have increased. Several people have even claimed to have photographed this "Big Foot," or Sasquatch, with

movie cameras, and some (not all) of these are very convincing. Some people claim these creatures could be a link between man and ape.

Let us assume that Big Foot is real. Would his existence prove the evolutionary "theory"? How else could such a creature be explained? _____

_____

_____

_____

_____

4. Botanists place the mosses and the flowering plants in separate phyla. This is because mosses have no organized vascular system (xylem and phloem) and no seeds, whereas flowering plants have phloem, xylem, flowers, and seeds. Many biologists believe that these two groups arose independently from different algal ancestors. Yet both mosses and flowering plants have guard cells which pull apart to form stomate pores. (See Exercise 8-3.) If these two phyla are derived from separate ancestors, how is it that they have the same complex cell structure of guard cells? What is meant by convergent evolution? Can this phenomenon be explained by something other than convergent evolution? _____

_____

_____

_____

_____

_____

5. In the arid Southwest of the United States there is an intriguing relationship between a moth and a plant. The plant is one of the yucca plants (a wide group of plants, a common one of which is known as the Spanish dagger), and it blooms only at night. When it blooms, the Yucca moth is attracted to it and pollinates it. The relationship does not stop there, however. The male and female moths will mate only inside the yucca flower, and then the female will lay her eggs only within the yucca plant itself. The plant cannot reproduce (and hence cannot survive as a species) without the moth. The moth in turn cannot reproduce without the yucca plant. These two species are mutually dependent and mutually beneficial. Try to arrive at a plan that would allow for this relationship to "evolve" over a long period of time through gradual mutations. Explain your conclusions. _____

_____

_____

_____

_____

_____

_____

_____

6. The above situation is called "mutualism."[19] Try to find other similar examples in nature. List as many as you can. _____

_____

_____

_____

7. Evolutionists often speak of "homology" as evidence of common ancestry. Homology means similarity of structures that serve different functions, such as the forelimb of the bat, the arm of a man, and the flipper of a dolphin. Are these similarities proof for evolution? Could they instead reflect a similarity of plan (e.g., art experts can look at a new painting and know who painted it because of certain unmistakable similarities between this painting and the artist's previous known works)? Explain. _____

_____

_____

_____

_____

_____

8. In Australia, almost all naturally occurring mammals are so-called "primitive" mammals and belong to the order Marsupialia. Evolutionists have explained this by pointing out that the evolutionary process is occurring more slowly in Australia than in other continents and that there has not been enough time for the evolution of more complex, placental forms. What might be some explanations for this lack of "higher" mammals from the creationist viewpoint? What problems relating to this might creationists pose for evolutionists? _____

_____

_____

_____

9. Evolutionists often say that the fossil evidence in various geological rock strata supports the "theory" of evolution and disproves creation. However, fossils of well-developed, "higher" forms of life have been found in the earliest, oldest Cambrian rocks. To date, the oldest skull of prehistoric man that has been found has been that of a "modern-type" human, according to the anthropologist who found it. Remains of modern-type man have been found in strata below the remains of Neanderthal man and other "primitive" ancestors. How might this evidence support the creation "theory" rather than the "theory of evolution"? _____

_____

_____

_____

19. *Mutualism* is the intimate living together of two dissimilar organisms in a mutually beneficial relationship; also known as *symbiosis*.

10. The question of the age of the earth has long been a problem for scientists and for theologians. The definition of the Hebrew word "yom" as used in the Bible has been interpreted to mean "day," as well as "period of time." Dating the earth by using genealogies has caused problems. The accuracy of the C^{14} method of dating has been questioned as has the use of other radioactive methods. What are some problems that both the theologian and the scientist must face when they try to date the earth's origin by biblical or scientific means? _____

    _____

    _____

    What are some common assumptions that are often made in these methods? _____

    _____

    _____

11. In Genesis 1, the author refers to reproduction of animals and plants "after their kind." Some scientists have tried to equate the word "kind" with the word "species." What problems arise with the definition of the term "species"?

    People have said that it is structurally impossible for two of every "kind" of animal to enter Noah's ark. How might your understanding of the words "kind" and "species" relate to an explanation to this statement? _____

    _____

    _____

12. Scientists often face the problem of why fossils of tropical plants and animals are found under the ice caps of the frozen polar regions. What evidence is found in the Bible which might offer an explanation for the presence of these life forms in these areas?

    _____

    Is it possible that at one time these regions had a different climate from that which they have today? _____

    _____

    _____

EXERCISE 9-4
VARIABILITY

INTRODUCTION:

All human beings look somewhat alike, and yet careful observation will show that variation exists in many categories. We vary in our height, weight, eye color, heart rates, and so forth. Evolutionists look on variation in the entire range of living organisms as the raw material through which new species of organisms can be produced. Creationists view variety as a marvelous mechanism by which man can select characteristics to meet his needs from a variety of characteristics already present in populations of organisms.

PROBLEM:

To study the patterns of variation in living things.

MATERIALS:

100 grasshoppers; 100 kidney beans; metric rulers.

PROCEDURE:

A. Variation in Grasshoppers and Kidney Beans

Remove the right jumping leg from each grasshopper. Extend the leg to its full length and measure it to the closest millimeter. Record your data in a table. For each different length determine the number of legs that have that length. Group your data in 12 equal intervals and then make a graph of your grouped data. Repeat for kidney beans.

Length of Leg	Number of Legs

Grouped Data	
Length	No.

Graph:

No. |
Lengths

Length of Bean	Number of Beans

Grouped Data	
Length	No.

Graph:

No. |
Lengths

B. Variation in Man

Obtain the pulse rate, height, and weight of twelve friends or adults all about the same age. Record your findings in the chart below. Group and graph your data on a separate sheet of paper.

Names of Participants	Pulse Rate	Height	Weight
1.			
2.			
3.			
4.			
5.			
6.			
7.			
8.			
9.			
10.			
11.			
12.			

DISCUSSION:

1. Look up the normal curve and compare it to your graph. Do some of your graphs come closer to the normal curve than others? _____ Explain why this may be so. _____

   _____

   _____

   _____

2. Some scientists believe that variation in organisms leads to the development of new species. Look at pictures of the races of man and read about "Darwin's Finches." From the creationist point of view, describe how the finches and races of man could have developed. _____

   _____

   _____

   How would you decide whether these forms were species or varieties? _____

   _____

EXERCISE 10-1
MAKING A LEAF COLLECTION

INTRODUCTION:

Leaf collections are commonly made by biology students to aid in identification of the common trees of their area, or as part of a habitat analysis. Making a leaf collection can be much fun, and since different types of trees grow in different soils and moisture conditions, a leaf collection from a particular location can tell you much about the ecology of that area.

PROBLEM:

To collect, preserve, and identify leaves from trees of your area.

MATERIALS:

Plant press; newspapers; corrugated cardboard; mounting paper; rubber cement; tree identification books.

PROCEDURE:

A leaf collection, like any other meaningful collection, involves three phases: collecting, preserving, and identifying the leaf. Collecting is no problem; all you have to do is to find different kinds of trees. *Be careful, however, that you do not violate the right of private property* to collect leaves; always secure permission to collect leaves from someone's yard. Although removal of a *few* leaves from a tree will not harm it, and most people are willing to cooperate with your collection, it is courteous and proper to ask first.

The problem of preservation of the leaves begins immediately after you remove them from the tree. Leaves must be kept flat, and they must be placed in something in which they will dry properly. If you are going out for only a couple of hours, a magazine is useful for temporary storage of your leaves. Place leaves from each different kind of tree between separate pages of the magazine, and make a notation on the edge of the page where and when the leaf was collected. Avoid using magazines with glossy pages, as the paper is not as absorbent and will delay the drying of your leaves.

When you get home, the leaves must be removed from the magazine and placed in a "press." (If you are on a long collection trip, place your leaves directly into the press.) The press is a simple device consisting of several layers of newsprint on both sides of each leaf group, then a layer of corrugated cardboard, another layer of newsprint and leaves, another layer of cardboard, and so forth. The entire bundle should be enclosed with stiff outer layers such as boards and then placed under a heavy weight or tied off tightly with a strap.

If there is much moisture in the leaves, you may have to change the newspaper several times to prevent mold. Normally, the leaves should remain in the press from one to two weeks, and should be checked occasionally to make sure that they are drying properly.

When the leaves are well-pressed and dried out, mount them on a sturdy paper with rubber cement. (Special plant-mounting paper is available, but drawing paper or white construction paper may be used.) Place leaves from only one tree on each page, and never use the back of a page. Try to display several leaves on each page, showing both the upper and lower leaf surface.

The labeling of your leaves should always include at least the scientific name of the plant, the date and location where it was collected, and the collector's name. You may wish to add the common name, the family, the altitude, habitat type, or other additional information.

Identification of the tree from which the leaf came is often left until last; however, at times this is not wise. Many students find that they have 5 sets of leaves from the same species of tree, or they find that they do not have an entire leaf (only part of a compound leaf). At least some partial identification should be attempted in the field as you collect. There are several good books available on tree identification. Check your library or book store, or ask your teacher to suggest a book or two. Following the exercise in identification of leaf parts below are some additional aids that will help you to tell the differences between the leaves more easily.

LEAF PARTS (identify)

Petiole _____

Midrib _____

Base of Blade _____

Margin of Blade _____

Blade _____

Apex of Blade _____

Veins _____

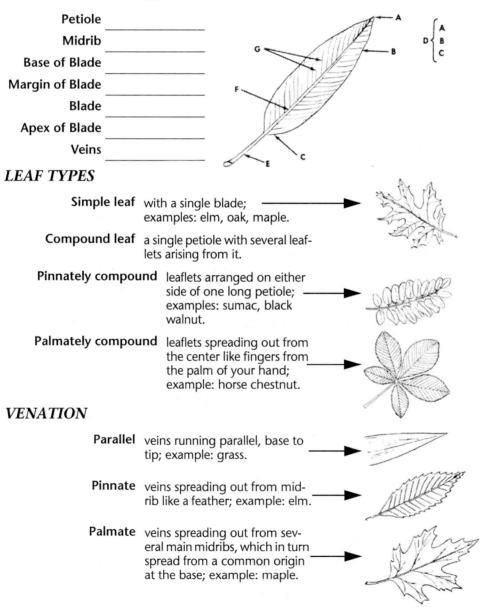

LEAF TYPES

Simple leaf   with a single blade; examples: elm, oak, maple.

Compound leaf   a single petiole with several leaflets arising from it.

Pinnately compound   leaflets arranged on either side of one long petiole; examples: sumac, black walnut.

Palmately compound   leaflets spreading out from the center like fingers from the palm of your hand; example: horse chestnut.

VENATION

Parallel   veins running parallel, base to tip; example: grass.

Pinnate   veins spreading out from midrib like a feather; example: elm.

Palmate   veins spreading out from several main midribs, which in turn spread from a common origin at the base; example: maple.

LEAF MARGINS

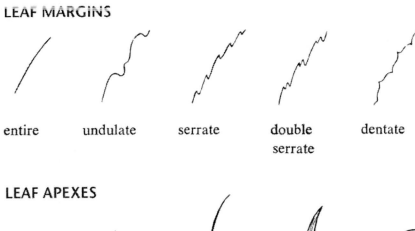

entire undulate serrate double serrate dentate lobed cleft

LEAF APEXES

acute obtuse aristate cuspidate rounded emarginate

LEAF BASES

acute rounded cordate uneven flattened saggitate

EXERCISE 10-2
MAKING AN INSECT COLLECTION

INTRODUCTION:

There are probably no animals that have more of an effect on our lives than insects. More than two thirds of all animal species are insects. Insects of one type or other live in almost every possible available habitat; underground, on the ground, in the tree tops, in the air, under the water, and even on the Antarctic ice cap. Most of these insects affect us in some way, either beneficially or harmfully. Making a collection of insects is a good way to learn more about the many kinds that live in your area, and may also be part of a habitat analysis.

PROBLEM:

To collect, preserve, and identify insects from your area.

MATERIALS:

Insect net; collecting jars; killing jars; ethyl acetate or carbon tetrachloride; insect pins (size 2 or 3); pinning boards; display boxes; insect identification books; (optional: aspirator; Berlese funnel).

PROCEDURE:

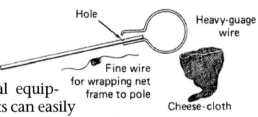

Collecting may be done in many ways. Sometimes you can merely pick insects up with your hands, but usually you will need some special equipment. An insect net for catching the flying insects can easily be made from an old broomstick handle, some cheesecloth, some heavy wire (approximately 20-gauge), and some fine wire (*see the figure above*). Many insects are attracted to street lamps or porch lights at night, and may be collected easily there. Some crawling insects can be coaxed right into your collecting bottle, but for smaller ones an aspirator works better (*see the figure at the right*). It is used to suck the insects into a jar; it works especially well on ants. The Berlese funnel (discussed in exercise 10-4) is another good device for collecting soil-living insects. Do not overlook turning over logs and rocks and examining the surface of plants.

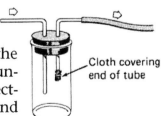

Before the insects can be mounted for your collection, they must be killed. Many entomologists use a killing jar containing a cyanide compound, but this can be very dangerous if you are not careful. Ethyl acetate is better, and is relatively safe. Plaster of paris, poured into a jar and allowed to dry, may be saturated with ethyl acetate. If the latter is not available, try carbon tetrachloride, which was once widely used as a spot remover. Carbon tetrachloride can be put on a piece of cloth in the bottom of your killing jar. Be sure to prepare your jar in a well-ventilated place, as this chemical can be harmful to you, also. Do not breathe the fumes. Specimens that are not easily killed by carbon tetrachloride or ethyl acetate may be killed by freezing them in your freezer. This is often done with large beetles.

Once your insect is dead, mount it as quickly as possible before it becomes brittle. Insect pins of size 2 or 3 are most preferred for pinning the insects. The pin should pass through the thorax, just off to one side (center it on flying insects). Push the pin in deeply, so that only a section big enough to hold is left above the insect. Put the insect on a pinning board and arrange the legs and antennae in a natural position, probing them in place with other pins. Insects such as butterflies, moths, and dragonflies should be mounted with their wings spread. Use a special spreading board to do this (*see the figure at the right*). Some people prefer to prepare their butterflies without a pin, and place them in a Riker mount,[20] which is a shallow box with poly-fill on the bottom and a glass top above the insect. When your pinned insects are dry, they should be pinned to a piece of cork in the bottom of a sturdy box. Labels giving the family, collector's name, date and location of collection are often pinned under the insect itself.

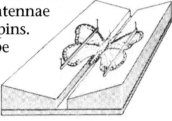

Spreading Board

Identification of insects requires a good field book. Many field books are available in your public or school libraries and local book stores. For most purposes, identification of the family name of the insect and the common name (if there is one) is usually sufficient.

---

20. A *Riker mount* is a display case filled with white polyfill batting; also called a *butterfly box*.

EXERCISE 10-3
SUCCESSION IN A FRESHWATER CULTURE

INTRODUCTION:

Living organisms are constantly undergoing change. They are born; they grow, mature, and then die. The same is true of living systems larger than the organism. For example, populations of certain organisms fluctuate or may even become extinct. The biological community, or biome (the total of all plant and animal populations plus the environmental factors), also has a life history full of change. Many people never recognize this change because it may take hundreds of years. However, in a freshwater pond, changes in the community (succession) occur in days or weeks. In this lab activity we will determine what environmental and population changes occur during succession in a freshwater pond.

PROBLEM:

To determine the environmental and population changes that may occur in a freshwater pond.

MATERIALS:

Boiled pond water; quart jars; medicine droppers; slides; cover slips; microscopes; dried plant material from the margins of lake, pond, or stream; pond organism identification manuals; hydrion paper (the cation state of hydrogen).

PROCEDURE:

A. Into a clean quart jar place a handful of the dried plant material and fill the jar with boiled pond water. Place the jar in moderate light.

B. Every other day for up to 3 weeks, study your pond community to determine the environmental conditions and the nature of the plant and animal populations. Record the temperature, measure and record the pH with hydrion paper, and indicate if water is clear or cloudy (turbid). Examine 5 drops of water from the surface, 5 drops from 5 cm down from the surface, and 5 drops from the bottom of the culture. Determine the average number of each organism for the 5 locations in that drop, and then for all 5 drops at each level. Use a pictorial guide to pond organisms or protozoans to aid in your identification of the organisms that you find. Record all your measurements and data in the following chart:

| | | | | | Average Number Per View | | |
Date	Temperature	pH	Turbidity	Organism	at Surface	at 5cm	at Bottom

(continue on an extra sheet if necessary)

163

After 3 weeks, make a graph to show the changes in temperature, one to show the pH, and a third one to show the turbidity changes. On each chart indicate "days" on the horizontal axis and the other factor on the vertical axis. If you can find several organisms that seem to be present most of the time, but in varying amounts, make a graph showing the relative changes in their populations. (Use those organisms that you list in questions 1 to 3.)

DISCUSSION:

1. What organism was most abundant during the first week? _____

   _____

2. What organism was most abundant during the second week? _____

   _____

3. What organism was most abundant during the third week? _____

   _____

4. What may have caused the organisms you observed during the first week to decline in

   their numbers? _____

   _____

   _____

5. Were any environmental changes accompanied by population changes? _____

   If so, provide possible explanations. _____

   _____

   _____

6. Describe a possible food chain or food web that may exist in your pond community.

   _____

   _____

   _____

7. Describe how you would decide whether or not your culture has reached a climax con-

   dition. _____

   _____

EXERCISE 10-4
SOILS AND SOIL PRODUCTIVITY

INTRODUCTION:

Except for rare instances, the plants that man depends upon most for his survival are those that grow with their roots in the soil. But not all soils are alike, and therefore plants and animals are also varied. Some soils are considered more fertile than others. In this lab activity we will compare various soil types and the organisms found in the soil to see whether this gives any clue to variations in soil productivity.

PROBLEM:

To determine the composition of various soils and to compare the soil organisms found in each.

MATERIALS:

3 samples each of soil (several different types); incubator (or oven); 500-ml graduated cylinder; Berlese funnel[21]; alcohol; high intensity lamp; cornmeal agar media; potato dextrose agar media; sterile water; transfer loops; slides; cover slips; 100-ml beakers; 100-ml graduated cylinders.

PROCEDURE:

Several different soil types should be available, but student teams should work with only one type of soil, and then share their observations.

A. Soil Composition

1. Weigh one soil sample and record its weight (use accurate measurement).

2. Put this soil sample into an incubator at 100°F for 24 hours. Weigh it again and record its weight on the chart on the next page. Compute the percentage of water content (water lost while heating).

3. Place another sample of the soil in a 500-ml graduated cylinder. Fill the cylinder with water, shake, and allow to settle. On the observation sheet, record the relative percentage of the volume of clay, sand, and humus in the soil.

B. Soil Organisms

1. Separate a third sample of one type of soil into 3 parts. From one part remove all visible animals. Place them in a small dish of alcohol. Identify the various kinds of animals found and their relative number.

2. Place the second part of the soil into a large funnel (you can make one with a roll of poster board if necessary). Cover the soil with wire screening and place a high intensity lamp over the soil. Place a small beaker of alcohol beneath the stem of the funnel. In

---

21. A Berlese funnel is a device in which soil is placed, and heat and light are applied from above, forcing mites, springtails (order Collembola), etc., into a container below it. To make your own funnel, see page 167.

24 hours examine the contents of the beaker. Separate the organisms by kind and indicate relative numbers.

3. From the third part of the soil sample, streak some soil on the surface of cornmeal agar in a petri dish. Also add 1 gr. of soil to 99 ml of sterile water. Shake well. Transfer 1 ml of this suspension to a petri dish of potato dextrose agar. Swirl the dish to spread the suspension. Incubate both cultures for a week and observe both with the naked eye and with the stereoscope. Make slides of the colonies for microscopic examinations. Describe the kinds and relative abundance of organisms on the table below.

TEST	Sample A	Sample B	Sample C
1. Weight of Soil Sample: before drying after drying Percent water (loss)			
2. Soil Composition: % volume humus % volume clay % volume sand			
3. Soil Organisms: large small (Berlese funnel) microscopic			

DISCUSSION:

What relationships exist between the type of soil and the kinds of organisms found in the soil sample? _____

_____

_____

_____

_____

_____

To Make a Berlese Funnel

MATERIALS:

A one-gallon plastic milk container (empty); an empty jelly jar (or a one-pint Mason jar) with a tight lid; a stick—about 25 cm long; 1/4" mesh hardware cloth or aluminum window screen (15 x 15 cm); a pair of scissors; masking tape or duct tape; rubbing alcohol (ethyl)—available at drug stores; a lamp.

PROCEDURE:

1. Cut the bottom out of the milk jug (*see figure at right*) and turn it upside down over the Mason jar to make a funnel.

2. Tape the stick to the handle of the milk jug (*see figure below*) so it is just long enough to reach the outside bottom of the Mason jar.

3. Bend down the corners of the hardware cloth so it fits snugly inside the wide end of the funnel. If using window screen, cut and pinch numerous slits so larger animals can crawl through.

4. Collect several handfuls of humus or leaf litter and put them on top of the wire mesh.

5. Pour alcohol into the Mason jar to a depth of 1 to 2 cm.

6. Carefully set the funnel on top of the jar and tape the stick to the jar so it won't tip over.

7. Leave the funnel in a warm, quiet place where it will not be disturbed.

8. Set a lamp over the funnel to speed drying. Keep the light bulb at least 10 cm away from the funnel.

MILK

Cut here

25 Watts

MILK

Tape stick to jar and milk jug.

MASON

Complete Setup

As the sample dries out, the animals will move down and fall into the alcohol. After 4 or 5 days (maybe longer if the sample was quite wet), you can *carefully* remove the jar and screw on its lid.   The alcohol will preserve the sample indefinitely.

EXERCISE 10-5
SAMPLING TERRESTRIAL ENVIRONMENTS

INTRODUCTION:

Much of biology can never be learned in a laboratory or from a textbook. In fact, until the nineteenth century, most of biology was field work. With the emphasis on ecology in the last fifty years, more and more workers are conducting biological studies of prairie fields and mountain forests. The best thing about field biology is that, to a limited extent, anyone can do original research. A very excellent and exciting way to gain understanding and appreciation of ecology is to sample and compare various types of habitats that are found in your area.

PROBLEM:

To sample the organisms present in several environments and to compare these with each other to find ecological relationships.

MATERIALS:

Identification keys for trees, shrubs, flowers, insects, animal tracks, birds, reptiles, and amphibians; traps (mouse traps or live traps); density boards; tree borers; soil-testing kits; materials used in lab exercises 10-1, 10-2, and 10-4; binoculars.

PROCEDURE:

This lab can best be done on a full-day or half-day field trip, with follow-up work done later in the classroom. If areas to be studied are close to the school, it might be done over a period of a week or so, using only the time allotted for lab periods. The class should study at least two different types of areas, and all areas should be studied as close to each other in time as possible so that seasonal variations do not confuse the comparison.

The sampling should be done for four basic types of observations: (1) plant diversity and density, (2) insect diversity, (3) diversity of other animals, and (4) soil evaluation. The class should be divided into four groups, and each group will make the same measurements in all areas studied. After all data are compiled by each group, the entire class can share their information and discuss their findings.

The types of measurements given are generally very meaningful, but under certain situations (in certain parts of the country) you may wish to delete certain measurements or add others that fit the study of your area better. The important thing is to obtain a wide range of measurements from several different habitats.

A. Plant Diversity and Density

A chart should be made in advance, giving space for plant types, name of plant, and relative abundance of it in your area (*see chart below*). Under plant type you may include: tree, shrub, and herb (or add others). Under each of these types, you should try to identify each plant. For trees this should be no problem (*see Exercise 10-1*). Any other plants that you know, or feel you can identify, you should name also. Many of the smaller plants may be very hard to identify, and it is not absolutely necessary to do this. For those plants that you cannot identify, simply collect a sample, press it, mount it, and give an identifying symbol (herb A, shrub B1, etc.). The important thing is to locate each different kind of plant and to determine its relative abundance. Below is a sample chart.

Plant Type	Plant Identification	Relative Abundance
Trees	sugar maple	common
	silver maple	uncommon
	black oak	rare
	red oak	abundant
	beech	dominant
Shrubs	gooseberry	common
	blackberry	uncommon
	shrub A	common near edges, uncommon elsewhere
	sumac	abundant
Herbs	grass A	uncommon
	bloodroot	common
	buttercup	common
	herb B	abundant—dominant

The type and relative abundance of plants present does not always tell the whole story of the habitat. You might have two plots, both having a particular plant that is relatively abundant, such as sugar maple trees. However, the one area might have all mature trees, towering high and shading the area below them, while the other might have only saplings with a fairly open canopy. The types of animals that you would expect to find in each area could be quite different.

One way to measure the density of cover in the different layers of a plot is with the "density board." For this, a six-foot-long board (4 to 6 inches wide) is painted with alternate foot-long sections of black and white paint. The board is painted with large numbers, starting with "1" at the bottom, proceeding up to "6" at the top. When it is put in place vertically in the field, an observer stands sixty to seventy feet away (standard is 66 feet) and reads the numbers that are unobscured. You then add all the numbers together for a score. Where there is no cover at all, all numbers are visible, and the score is 21 (1 + 2 + 3 + 4 + 5 + 6 = 21). If only number 3 and 4 are visible, the score is 7, and that would indicate a fairly dense cover. This measurement should be taken at least 5 to 10 times in your area and then averaged. Record both the numbers that were visible and the total score.

It might also be meaningful to get an average-size measurement of some of the trees or shrubs present. An entire plot of trees averaging 2 to 4 inches thick is much different from a plot where most trees are a foot or more in diameter. If you have a tree borer, you may even wish to take a core from several trees and try to determine their age by counting the rings.

Another method for obtaining a relative density measurement is to take a reading with a light meter (such as from a camera) at various heights (beginning at 2 inches, then every foot level from 1 to 6) in each of your study areas. Compare this with the light available by dividing your readings by the reading in full sunlight to obtain a percentage of light penetration. Repeat in several locations at random throughout the study area.

B. Insect Diversity

This is a fairly simple measure, but it may take considerable time to do a good job of it. Attempt to collect as many different kinds of insects and other arthropods in the area, and maintain a record of relative abundance of each one. You may also wish to note where each insect was found, such as on a sunflower, under a log, etc., and whether it was at the edge of the woods, in the middle of a field, etc.

The more information that you can include, the better your comparisons will be. Follow collection procedures given in Exercise 10-2. Be sure to sample every possible part of the environment (including treetops, just above the grass or bushes, [sweep this area with your net], logs or bark, and on plants).

C. Diversity of Other Animals

This study might seem almost too easy, but to do a really good job requires hard work, patience, and skill. Many people live in an area or even hunt and fish in an area all of their lives, and still do not see all of the animals that there are in it. Your skill as a woodsman and a scientist will be tested to its fullest here. Make a list of birds seen (or heard), nests or dens found (some dens can be identified by smell or shape), reptiles and amphibians and mammals seen. Mammals are much harder to detect in an area, especially when there are many people in it.

You will see some mammals such as squirrels and chipmunks, but for most others you will have to depend upon indirect evidence. Dens were already mentioned, but do not forget to look for tracks and droppings. A good field guide to animal tracks will help you identify tracks that you do not recognize. Owls often have regular perches in a tree where they sit to eat. Before leaving, they will regurgitate "pellets" containing bones and hair. Pulling these apart with dissecting needles and trying to identify the bone or hair is a very fruitful method of finding out not only what the owl eats, but also what small mammals are in the area. You might need some help to identify the bones and hair, but it is fun trying on your own.

Deer and rabbits have distinct ways of chewing on vegetation, and the amount of this type of destruction is an indication of their abundance also. Small mammals (mice, shrews, moles, etc.) can be trapped in mouse traps, or in a variety of live traps. Don't fail to overturn stones to look for snakes. (When doing this, always lift the side farthest from you, so that you do not expose your legs to a venomous snake that might be underneath it.) In final summary, detection of the presence of many of these other animals is like investigating a mystery: it takes a shrewd detective to put all the clues together, but it is very rewarding.

D. Soil Evaluation

It is not too surprising to discover that different kinds of plants and animals are found in relation to different types of soils. There are several measurements that can be made easily on soils.

A garden-soil testing kit such as those sold in garden stores can be used for several tests. They usually include tests for pH, and for absence of various minerals such as nitrogen, phosphates, or potassium. These are used by the gardener to tell him what type and how much fertilizer he needs. You can use them to reveal interesting discoveries about the differences in soils in your study areas.

Refer to Exercise 10-4 for other techniques of soil sampling. Be sure to test random samples from each area (at least 3 to 5 samples) for water content, organic content (by baking out the humus), digging down to determine the depth of the top soil (in a forest also measure the depth of the "litter," or the partially decomposed leaves on the forest floor). Check both the leaf litter and the soil itself for life by using the Berlese funnel as described in Exercise 10-4.

COMPARISON OF DATA:

When all groups have finished with each area of study, you should spend a considerable amount of time compiling your data and comparing the results from each different area. Members of each group should share their discoveries with the class by giving a report of what they found.

DISCUSSION:

1. What characteristics of the different areas seemed least obviously different? _____

   _____

   _____

   Can you give any reason for this? _____

   _____

   _____

2. Can you select one or two characteristics of each area that clearly distinguish it from the other areas? _____

   _____

3. What differences would you expect to find in each area if you had seen it fifty years ago? _____

   _____

   _____

   _____

4. What differences would you expect to find in each area if you were to see it fifty years from now? _____

   _____

   _____

   _____

EXERCISE 10-6
SAMPLING AQUATIC ENVIRONMENTS

INTRODUCTION:

Bodies of water support more life than all other habitats. Not only does water cover most of the earth's surface, but organisms can also exist at different depths from surface to bottom. Bodies of water often undergo changes from season to season and from year to year.

When lakes are first formed by volcanoes, glaciers, or other means, they are usually classified as *oligotrophic* (ä'lə•gä•trō'fĭk, deficient in plant nutrients), meaning that they are quite deep, cold, and relatively unproductive. As time progresses, the shorelines wear down and life becomes more abundant. As organisms die and settle to the bottom to decompose, and as silt and dirt are brought into the lake from incoming streams, the lake becomes more shallow. Now the lake is known as a eutrophic lake because it is quite productive, warmer, and shallower.

Pollution usually speeds up this process of succession known as eutrophication because it adds mineral nutrients to the water, causing plants and animals to multiply faster. Although the study of these bodies of water is often difficult, many interesting plants and animals can be found, and many important ecological processes can be observed. Almost any established body of water, whether it be a roadside ditch or an ocean, offers abundant life forms.

PROBLEM:

1. To observe the stages of succession in aquatic communities and to recognize their various life zones.

2. To sample the organisms in different aquatic environments and to make comparisons which indicate differences in habitat and in behavior.

MATERIALS:

Picture keys or dichotomous keys for most aquatic plants and animals; colored pencils; small mesh minnow seine; collecting jars; 10% formalin solution (formaldehyde containing a small amount of methanol); specimen tags or labels; slides and cover slips; microscopes.

PROCEDURE:

A. Life Zones of a Lake

Visit a lake, preferably a small lake where you can see the entire shoreline from one point. Look at the surrounding vegetation and notice the different life zones. You will probably see something like what is pictured on the next page:

Zone 1	a climax forest of tall trees farthest from shore
Zone 2	smaller trees, shrubs, and bushes nearer the water
Zone 3	grasses and other soft-stemmed plants
Zone 4	plants rooted in the water but whose stems emerge from the water
Zone 5	plants rooted in water whose leaves float on the surface
Zone 6	rooted plants that are entirely submerged
Zone 7	open water zone (Only free-swimming organisms live here.)

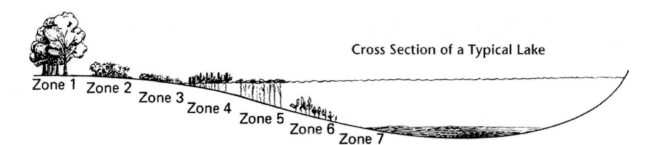

CROSS SECTION OF A TYPICAL LAKE

Draw below a sketch of the lake you are studying. Using different colored pencils, indicate the different life zones and show the relative amount of space each zone covers.

Which of the life zones in the lake you are studying covers the most area? _____

How does the distance from shore of Zone 1 (the climax forest) indicate how rapidly succession is taking place? _____

_____

What will eventually happen to a lake if the process of succession continues for many years or possibly many hundreds of years? _____

_____

_____

What two factors determine how deep in the water the free-swimming plants and animals in Zone 7 can exist? _____

_____

B. **Sampling and Comparing Aquatic Organisms**

Collect some water from the lake near the edge. With the water, take a few small plants and a little mud or sand from the bottom. Be sure to keep air supplied to the water as sealing the container tightly may result in the suffocation of any living organisms. Take this back to the laboratory and with a microscope look for protozoans, diatoms, and other protists. How do these organisms play an important role in the aquatic community? _____

_____

Examine the water and land near the shoreline and notice the various forms of any plants and animals. Using the same procedure outlined in Exercise 10-5, make a list of any plants and animals that you can recognize and indicate their relative abundance. Use a key to identify those that you do not know. Examples of abundant plants may include duckweed, cattails, water lilies, arrow weed, algae, and so forth. Common animals may include hydra, planaria, insects or insect larvae, crayfish, frogs and salamanders, clams and snails, turtles and snakes.

Using a minnow seine, capture a few specimens of as many different species of fish as you can. Place these in a collecting jar with enough 10% formalin solution to cover them. Also place in the jar with the fish a specimen tag or label on which you have written in pencil the location of capture and other collection data. These are to be separated and identified later in the laboratory. Now go to a completely different type of aquatic environment such as a fast-moving stream and again collect as many different species of fish as possible. Preserve these in the same way using a different container, label them, and take them to the laboratory for identification.

In the laboratory, use a picture key or dichotomous key and separate the fish from the lake into groups by species and identify them. Place these on one side of the laboratory. On the other side of the room, separate the fish from the stream into groups by species.

Now make comparisons:

Examine the fish for presence of teeth. How might this indicate their diet? _____

_____

What are some differences in the number, size, and shape of the teeth? _____

_____

_____

Compare the fish captured in the lake with those captured in swift water. In general, which group has members with the longest, most streamlined body? _____

_____

Considering body shape and morphology alone, why would it be difficult for many lake fish to exist in fast-moving streams? _____

_____

Look on the head of certain fish for fleshy, whiskerlike structures known as barbels. On which of the two groups of fish do you sometimes find these? _____
What is their function? _____

_____

Compare the mouth position of the different species. Look for a ventrally located sucker-type mouth on some fish. If such fish are available, study the structure of the jaws. Where do fish with this type of mouth get their food? _____

_____

In which of the two habitats did you collect these fish? _____

Some fish may have a mouth that is turned upward. Where do you suppose they feed, and what do they probably eat? _____

_____

Which fish would you expect to have the highest oxygen requirements, those in the shallow waters of a lake or those in a fast-moving stream? _____

_____

Make other comparisons such as the number of dorsal fins and their location on the body, the number of rays and spines in the fins, the shape of the caudal fin, the location of the pectoral and pelvic fins, and their color. How might these morphological differences indicate differences in behavior, diet, habitat, or life habits? _____

_____

_____

_____

_____

_____

_____

_____

_____

EXERCISE 10-7
PRODUCING A SELF-SUFFICIENT ENVIRONMENT

INTRODUCTION:

For years, the pioneers of early America considered the great wealth of natural resources to be unlimited. Today, vast areas of this "Promised Land" now are wasteland, and many animals that were once abundant are now extinct. We are gradually discovering that if we destroy one part of our environment, it affects the other parts, also. Our planet does not have an endless supply of materials; it is more like an island with only limited resources available. Ecologists talk about a balance in nature among the various plants and animals. Just how delicate is this balance? Is it more delicate in some areas than others? What happens when the balance is destroyed?

PROBLEM:

The above are basic questions that can be answered only on a large scale. On a smaller scale we might ask: What basic requirements are necessary to maintain a self-sufficient environment, and how does it operate? This exercise will demonstrate a simple balance and how it functions.

MATERIALS:

6 large test tubes; 6 rubber stoppers; candle; aquarium water; 5 sprigs of aquarium plants (elodea); 5 snails (small ones); a dark closet; a grow light or a desk lamp; bromthymol blue.[22]

PROCEDURE:

Add aquarium water to each of the six test tubes until they are about half full. Then prepare each of them with the following combinations:

1. elodea only

2. snail only

3. elodea and snail

4. elodea, snail, and a few drops of bromthymol blue turned yellow

5. elodea, snail, bromthymol blue as above

6. elodea, snail, bromthymol blue as above

    Plug each test tube tightly with a cork to prevent any exchange of air. To ensure a tight seal, drip candle wax around each rubber stopper. Place test tubes 1, 2, 3, and 4 in the room near a window; place test tube 5 in a dark closet or other continually dark place; and place test tube 6 under a special growing lamp (or a desk lamp) that is kept on con-

---

22. *Bromthymol blue* is a dye derived from thymol that is an acid-base indicator; *thymol* is a crystalline phenol $C_{10}H_{14}O$ of aromatic odor and antiseptic properties.

stantly 24 hours a day. Examine each test tube at least once a day and record the results each day, until no further changes are apparent. At least once, check test tubes 4 to 6 in the early morning and again in the late afternoon.

DISCUSSION:

1. Of what is bromthymol blue an indicator? _____

    _____

2. What color is the bromthymol blue at the beginning of the day (after a full night of darkness)? _____ Why? _____

    _____

    _____

3. What color is it later in the day? _____ Why? _____

    _____

    _____

4. Compare the results shown in the tube in the dark and the one under constant light with the results seen in the tubes kept under normal light cycle. Explain the differences. _____

    _____

    _____

    _____

5. Make a brief statement on what is involved in "balance" in this exercise. _____

    _____

    _____

6. Why might this exercise be considered as an oversimplification of the situation in nature? _____

    _____

7. We called this an experiment with a "self-sufficient environment." Is it truly self-sufficient? _____ Why or why not? _____

    _____

EXERCISE 10-8
DETERMINING THE DENSITY OF ANIMAL POPULATIONS

INTRODUCTION:

Often an ecologist finds he needs to know the number of animals in a particular population or in a given area. He then can determine what relationships exist in a community and how much competition there is within a population or between populations. Large animals can often be counted directly, or rough estimates can be made by counting droppings or tracks. Estimating population densities of smaller animals or of animals in the water is often a more difficult task.

In this lab activity we will consider one common method of determining population density by using a procedure called the Lincoln-Peterson Index. In the use of this method, animals are captured and permanently marked by placing bands on the legs of birds, by clipping a certain toenail of mice, or by placing tags in the dorsal fin of fish. These marked individuals are then returned to the population, and a number of animals are again captured. The ratio of marked to unmarked individuals in the second capture is determined, and the population density is calculated by using the following formula: **P:M = p:m** (P = total in the population; M = number of marked individuals in the population, p = number of individuals in the second capture; and m = number of marked individuals in the second capture).

PROBLEMS:

1. To learn how to estimate population density

2. To learn how to recognize assumptions and weaknesses in scientific methodology

MATERIALS:

Several thousand slips of white paper cut into 1/2-inch squares; small paper bags.

PROCEDURE:

Take a generous handful of paper slips and place them in a paper bag. About how many slips would you guess are in the bag? _____ Now remove some of the slips one by one and make a pencil mark on each side until you have a pile of about fifty marked slips. Return these marked slips to the bag and mix them thoroughly with the other slips. Now remove approximately fifty slips one by one. Of these fifty slips, how many have marks on them? _____ How many are unmarked? _____ Now fill in the formula and calculate how many total slips were originally in the bag.

P:M = p:m (_____:_____ = _____:_____) What is your estimate? _____

Return the unmarked slips to the bag and remove all marked slips from the bag so that you again have your original pile of marked slips. Now remove about fifty more slips, mark them on both sides, and return these along with your original marked pile to the paper bag with the unmarked slips. After mixing them thoroughly, remove approximately one hundred slips and determine the ratio of marked to unmarked slips. Fill in the formula again and calculate the original number of slips in the bag.

P:M = p:m (_____:_____ = _____:_____) What is your estimate? _____

Now do the same as above, this time using approximately 200 marked slips. Again calculate the number of slips in the bag.

P:M = p:m (_____:_____ = _____:_____) What is your estimate? _____

Which of the four estimates that you made do you expect to be the most accurate? _____ Why? _____

_____

Now make an actual count of all marked and unmarked slips. Which of the four estimates was the most accurate? _____

As with many tests, this procedure, applied to an actual animal population, involves several assumptions if it is to be considered accurate. What are some of these assumptions? _____

_____

_____

_____

_____

_____

_____